Noise Pollution and Control

Noise Pollution and Control

Edited by
Noel Templeton

Larsen & Keller
www.larsen-keller.com

Noise Pollution and Control
Edited by Noel Templeton
ISBN: 978-1-63549-197-5(Hardback)

Larsen & Keller
Published by Larsen and Keller Education,
5 Penn Plaza,
19th Floor,
New York, NY 10001, USA

Cataloging-in-Publication Data

Noise pollution and control / edited by Noel Templeton.
p. cm.
Includes bibliographical references and index.
ISBN 978-1-63549-197-5
1. Noise pollution. 2. Noise control. 3. Noise--Measurement.
4. Pollution prevention. I. Templeton, Noel.
TD892 .N65 2017
620.23--dc23

Printed and bound in China.

For more information regarding Larsen and Keller Education and its products, please visit the publisher's website www.larsen-keller.com

Table of Contents

Preface

The rise in vehicles, engines, motors, acoustic devices and other factors are causing noise pollution to increase at an alarming rate. Pollution is highly detrimental to our environment. Thus, its control and prevention are crucial. The book is designed to provide the students some fundamental concepts of noise pollution and its control. It contains chapters that deal with the causes of pollution and the various technologies to control it. The book studies, analyses and upholds the pillars of this subject and its utmost significance in modern times. It is compiled in such a manner that it will provide in-depth knowledge about the theory and practice of this field. This textbook is meant for students who are looking for an elaborate reference text on noise pollution and control.

To facilitate a deeper understanding of the contents of this book a short introduction of every chapter is written below:

Chapter 1- Excessive noise that causes discomfort and potential physical harm is known as noise pollution. Noise pollution is mainly caused by machines and by transportation systems. It is increasingly becoming a cause of concern for environmentalists all across the world as high levels of noise can result in the death of animals and also certain disabilities in humans.

Chapter 2- This chapter elucidates on certain examples of noise pollution such as noise caused by industries, roadway noise, jet noise, aircraft noise and environmental noise. Noise pollution is as important to deal with as air pollution or water pollution. Understanding noise pollution is necessary if we want to control it in time. Noise pollution is best understood in confluence with the major topics listed in the following chapter.

Chapter 3- To mark a difference between sound and noise we need some level of measurement. Sound exposure level and sound intensity level are two levels explained within this chapter. The following section helps in providing the reader with an in-depth understanding of the subject matter.

Chapter 4- Noise pollution has certain effects on the environment as well as an individual. Some of these effects are hearing problems, cardiovascular issues, sleeping disorders and tinnitus. Animals also face problems because of noise pollution as the frequency of their hearing is higher. The effects may not be felt strongly on a day-to-day basis, but it can be very damaging in the long run. This section helps the readers in understanding the effects produced by noise pollution.

Chapter 5- There are few technologies that can prevent noise pollution. Some of them are construction of soundproof rooms for noisy machines, absorption and vibration isolation. Excessive noise can affect psychological health, cause sleeping disorders and can also cause death in animals. This chapter has been carefully written to provide an easy understanding of noise control.

Chapter 6- Sound is the vibration that is audible and travels through a medium such as air or water. Elements such as pitch, duration, loudness, timbre and sound localization can analyze sound. In order to completely understand sound it is necessary to understand the properties related to it.

Chapter 7- Audiology is a branch of science that studies topics such as hearing and the disorders related to hearing. The aspects explained in the following section are audiometers, audiograms, audiometrists, audiometry and pure tone audiometry. This section is an overview of the subject matter incorporating all the main features of audiology.

I would like to share the credit of this book with my editorial team who worked tirelessly on this book. I owe the completion of this book to the never-ending support of my family, who supported me throughout the project.

Editor

1

Introduction to Noise Pollution

Excessive noise that causes discomfort and potential physical harm is known as noise pollution. Noise pollution is mainly caused by machines and by transportation systems. It is increasingly becoming a cause of concern for environmentalists all across the world as high levels of noise can result in the death of animals and also certain disabilities in humans.

Noise Pollution

Noise pollution or noise disturbance is the disturbing or excessive noise that may harm the activity or balance of human or animal life. The source of most outdoor noise worldwide is mainly caused by machines and transportation systems, motor vehicles, aircraft, and trains. Outdoor noise is summarized by the word environmental noise. Poor urban planning may give rise to noise pollution, since side-by-side industrial and residential buildings can result in noise pollution in the residential areas. Documented problems associated with urban noise go back as far as Ancient Rome.

Traffic is the main source of noise pollution in cities.

Outdoor noise can be caused by machines, construction activities, and music performances, especially in some workplaces. Noise-induced hearing loss can be caused by outside (e.g. trains) or inside (e.g. music) noise.

A Qantas Airways Boeing 747-400 passes close to houses shortly before landing at London Heathrow Airport.

High noise levels can contribute to cardiovascular effects in humans and an increased incidence of coronary artery disease. In animals, noise can increase the risk of death by altering predator or prey detection and avoidance, interfere with reproduction and navigation, and contribute to permanent hearing loss.

Health

Humans

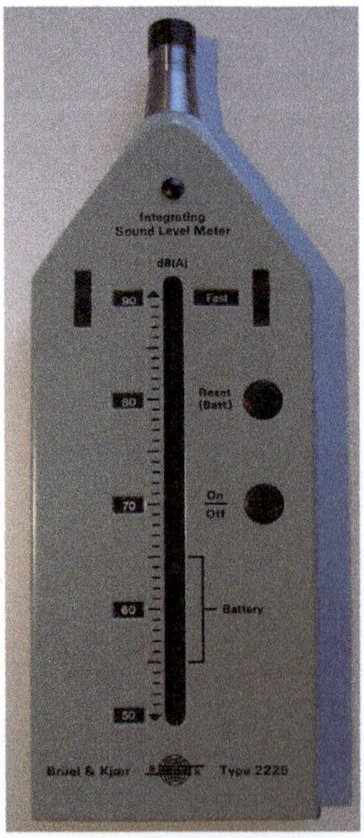

A sound level meter, a basic tool in measuring sound.

Noise pollution affects both health and behavior. Unwanted sound (noise) can damage psychological health. Noise pollution can cause hypertension, high stress levels, tinnitus, hearing loss, sleep disturbances, and other harmful effects.

Reaction to noise

Sound becomes unwanted when it either interferes with normal activities such as sleeping, conversation, or disrupts or diminishes one's quality of life.

Chronic exposure to noise may cause noise-induced hearing loss. Older males exposed to significant occupational noise demonstrate more significantly reduced hearing sensitivity than their non-exposed peers, though differences in hearing sensitivity decrease with time and the two groups are indistinguishable by age 79. A comparison of Maaban tribesmen, who were insignificantly exposed to transportation or industrial noise, to a typical U.S. population showed that chronic exposure to moderately high levels of environmental noise contributes to hearing loss.

High noise levels can result in cardiovascular effects and exposure to moderately high levels during a single eight-hour period causes a statistical rise in blood pressure of five to ten points and an increase in stress, and vasoconstriction leading to the increased blood pressure noted above, as well as to increased incidence of coronary artery disease.

Less addressed is how humans adapt to noise subjectively. Indeed, tolerance for noise is frequently independent of decibel levels. However, Murray Schafer's soundscape research was groundbreaking in this regard. In his eponymous work, he makes compelling arguments about how humans relate to noise on a subjective level, and how such subjectivity is conditioned by culture. He also notes that sound is an expression of power, and as such, material culture (e.g., fast cars or Harley Davidson motorcycles with aftermarket pipes) tend to have louder engines not only for safety reasons, but for expressions of power by dominating the soundscape with a particular sound. Other key research in this area can be seen in Fong's comparative analysis of soundscape differences between Bangkok, Thailand and Los Angeles, California, US. Fong's research methodology was modeled after Schafer, and the research findings show how not only do soundscapes differ, but they also rather explicitly point to the level of urban development in the area; that is, cities in the periphery - in Immanuel Wallerstein-speak - will have different soundscapes than that of cities in the core. Fong's important findings tie not only soundscape appreciation to our subjective views of sound, but also demonstrates how different sounds of the soundscape are indicative of class differences in urban environments.

Wildlife

Noise can have a detrimental effect on wild animals, increasing the risk of death by changing the delicate balance in predator or prey detection and avoidance, and interfering the use of the sounds in communication, especially in relation to reproduction and in navigation. Acoustic overexposure can lead to temporary or permanent loss of hearing.

An impact of noise on wild animal life is the reduction of usable habitat that noisy areas may cause, which in the case of endangered species may be part of the path to extinction. Noise pollution may have caused the death of certain species of whales that beached themselves after being exposed to the loud sound of military sonar.

Noise also makes species communicate more loudly, which is called Lombard vocal response. Scientists and researchers have conducted experiments that show whales' song length is longer when submarine-detectors are on. If creatures do not "speak" loudly enough, their voice will be masked by anthropogenic sounds. These unheard voices might be warnings, finding of prey, or preparations of net-bubbling. When one species begins speaking more loudly, it will mask other species' voice, causing the whole ecosystem eventually to speak more loudly.

Marine invertebrates, such as crabs (*Carcinus maenas*), have also been shown to be negatively affected by ship noise. Larger crabs were noted to be negatively affected more by the sounds than smaller crabs. Repeated exposure to the sounds did lead to acclimatization.

European robins living in urban environments are more likely to sing at night in places with high levels of noise pollution during the day, suggesting that they sing at night because it is quieter, and their message can propagate through the environment more clearly. The same study showed that daytime noise was a stronger predictor of nocturnal singing than night-time light pollution, to which the phenomenon often is attributed.

Zebra finches become less faithful to their partners when exposed to traffic noise. This could alter a population's evolutionary trajectory by selecting traits, sapping resources normally devoted to other activities and thus leading to profound genetic and evolutionary consequences.

Noise Mitigation

The sound tube in Melbourne, Australia is designed to reduce roadway noise without detracting from the area's aesthetics

Roadway noise can be reduced by the use of noise barriers, limitation of vehicle speeds, alteration of roadway surface texture, limitation of heavy vehicles, use of traffic controls that smooth vehicle flow to reduce braking and acceleration, and tire design. An important factor in applying these strategies is a computer model for roadway noise, that is capable of addressing local topography, meteorology, traffic operations, and hypothetical mitigation. Costs of building-in mitigation can be modest, provided these solutions are sought in the planning stage of a roadway project.

Aircraft noise can be reduced by using quieter jet engines. Altering flight paths and time of day runway has benefitted residents near airports.

Industrial noise has been addressed since the 1930s via redesign of industrial equipment, shock mounted assemblies and physical barriers in the workplace. In recent years, Buy Quiet programs and initiatives have arisen in an effort to combat occupational noise exposures. These programs promote the purchase of quieter tools and equipment and encourage manufacturers to design quieter equipment. The US National Institute for Occupational Health has created a database of industrial equipment with decibel levels noted.

Legal Status

Up until the 1970s governments tended to view noise as a "nuisance" rather than an environmental problem.

Many conflicts over noise pollution are handled by negotiation between the emitter and the receiver. Escalation procedures vary by country, and may include action in conjunction with local authorities, in particular the police.

India

Noise pollution is a major problem in India. The government of India has rules & regulations against firecrackers and loudspeakers, but enforcement is extremely lax. Awaaz (sound) Foundation is an Indian NGO working to control noise pollution from various sources through advocacy, public interest litigation, awareness, and educational campaigns since 2003.Despite increased enforcement and stringency of laws now being practised in urban areas, rural areas are still affected.

United Kingdom

Figures compiled by rockwool, the mineral wool insulation manufacturer, based on responses from local authorities to a Freedom of Information Act (FOI) request reveal in the period April 2008 – 2009 UK councils received 315,838 complaints about noise pollution from private residences. This resulted in environmental health officers across the UK serving 8,069 noise abatement notices or citations under the terms of the Anti-Social Behaviour (Scotland) Act. In the last 12 months, 524 confiscations of equipment have been authorized involving the removal of powerful speakers, stereos and televisions. Westminster City Council has received more complaints per head of population than any other district in the UK with 9,814 grievances about noise, which equates to 42.32 complaints per thousand residents. Eight of the top 10 councils ranked by complaints per 1,000 residents are located in London.

United States

There are federal standards for highway and aircraft noise; states and local governments typically have very specific statutes on building codes, urban planning, and roadway development.

Noise laws and ordinances vary widely among municipalities and indeed do not even exist in some cities. An ordinance may contain a general prohibition against making noise that is a nuisance, or it may set out specific guidelines for the level of noise allowable at certain times of the day and for certain activities.

The Environmental Protection Agency retains authority to investigate and study noise and its effect, disseminate information to the public regarding noise pollution and its adverse health effects, respond to inquiries on matters related to noise, and evaluate the effectiveness of existing regulations for protecting the public health and welfare, pursuant to the Noise Control Act of 1972 and the Quiet Communities Act of 1978.

New York City instituted the first comprehensive noise code in 1985. The Portland Noise Code includes potential fines of up to $5000 per infraction and is the basis for other major U.S. and Canadian city noise ordinances.

Noise

Noise is unwanted sound judged to be unpleasant, loud or disruptive to hearing. From a physics standpoint, noise is indistinguishable from sound, as both are vibrations through a medium, such as air or water. The difference arises when the brain receives and perceives a sound.

NASA researchers at Glenn Research Center conducting tests on aircraft engine noise in 1967

In experimental sciences, noise can refer to any random fluctuations of data that hinders perception of an expected signal.

Acoustic noise is any sound in the acoustic domain, either deliberate (e.g., music or speech) or unintended. In contrast, noise in electronics may not be audible to the human ear and may require instruments for detection.

In audio engineering, noise can refer to the unwanted residual electronic noise signal that gives rise to acoustic noise heard as a hiss. This signal noise is commonly measured using A-weighting or ITU-R 468 weighting.

Measurement

Sound is measured based on the amplitude and frequency of a sound wave. Noise is most commonly discussed in terms of decibels (dB), the measure of loudness, or intensity of a sound; this measurement describes the amplitude of a sound wave. On the other hand, pitch describes the frequency of a sound and is measured in hertz (Hz).

Recording and Reproduction

In audio, recording, and broadcast systems, audio noise refers to the residual low-level sound (four major types: hiss, rumble, crackle, and hum) that is heard in quiet periods of program. This variation from the expected pure sound or silence can be caused by the audio recording equipment, the instrument, or ambient noise in the recording room.

In audio engineering it can refer either to the acoustic noise from loudspeakers or to the unwanted residual electronic noise signal that gives rise to acoustic noise heard as 'hiss'. This signal noise is commonly measured using A-weighting or ITU-R 468 weighting

Noise is often generated deliberately and used as a test signal for audio recording and reproduction equipment.

White Noise

White noise is energy randomly spread across a wide frequency band containing all notes from high to low. It is called "white" noise as it is analagous to "white" light which contains all the colors of the visible spectrum.

Environmental Noise

Environmental noise is the accumulation of all noise present in a specified environment. The principal sources of environmental noise are surface motor vehicles, aircraft, trains and industrial sources. These noise sources expose millions of people to noise pollution that creates not only annoyance, but also significant health consequences such as elevated incidence of hearing loss and cardiovascular disease. There are a variety of mitigation strategies and controls available to reduce sound levels including source intensity reduction, land use planning strategies, noise barriers and sound baffles, time of day use regimens, vehicle operational controls and architectural acoustics design measures.

Regulation of Noise

Certain geographic areas or specific occupations may be at a higher risk of being exposed to constantly high levels of noise; in order to prevent negative health outcomes, regulations may be set. Noise regulation includes statutes or guidelines relating to sound transmission established by national, state or provincial and municipal levels of government. Environmental noise is governed by laws and standards which set maximum recommended levels of noise for specific land uses, such as residential areas, areas of outstanding natural beauty, or schools. These standards usually specify measurement using a weighting filter, most often A-weighting.

United States

In 1972, the Noise Control Act was passed to promote a healthy living environment for all Americans, where noise does not pose a threat to human health. This policy's main objectives were: (1) establish coordination of research in the area of noise control, (2) establish federal standards on noise emission for commercial products, and (3) promote public awareness about noise emission and reduction.

The Quiet Communities Act of 1978 promotes noise control programs at the state and local level and developed a research program on noise control. Both laws authorized the Environmental Protection Agency to study the effects of noise and evaluate regulations regarding noise control.

Noise in the Workplace

In the US, the National Institute of Occupational Safety and Health (NIOSH) provides recommendation on noise exposure in the workplace. In 1972 (revised in 1998), NIOSH published a document outlining recommended standards relating to the occupational exposure to noise, with the purpose of reducing the risk of developing permanent hearing loss related to exposure at work. This publication set the recommended exposure limit (REL) of noise in an occupation setting to 85 dBA for 8 hours. However, in 1973 the Occupational Safety and Health Administration (OSHA) maintained the requirement of an 8-hour average of 90 dBA. The following year, OSHA required employers to provide a hearing conservation program to workers exposed to 85 dBA average 8-hour workdays.

Europe

The European Environment Agency regulates noise control and surveillance within the European Union. The Environmental Noise Directive was set to determine levels of noise exposure, increase public access to information regarding environmental noise, and reduce environmental noise. Additionally, in the European Union, underwater noise is a pollutant according to the Marine Strategy Framework Directive (MSFD). The MSFD requires EU Member States to achieve or maintain Good Environmental Status, meaning that the "introduction of energy, including underwater noise, is at levels that do not adversely affect the marine environment".

Health Effects from Noise

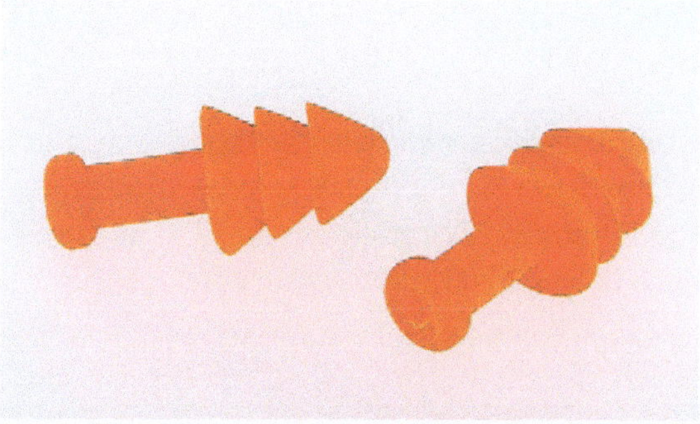

Earplugs can be used to protect the user's ears from loud noises.

Exposure to noise is associated with several negative health outcomes. Depending on duration and level of exposure, noise may promote hearing loss, high blood pressure, ischemic heart disease, sleep disturbances, birth defects, and even decreased school performance.

Noise exposure has increasingly been identified as a public health issue, especially in an occupational setting, as demonstrated with the creation of NIOSH's Noise and Hearing Loss Prevention program. Noise has also proven to be an occupational hazard, as it is the most common work-related illness.

Prevention

While noise-induced hearing loss is permanent, it is also very preventable. Particularly in the workplace, regulations may exist depicting maximum allowed levels of noise. This can be especially important for professionals working in atmospheres with consistent exposure to noise, such as musicians, music teachers and sound engineers. Examples of measures taken to prevent noise-induced hearing loss in the workplace include engineering noise control, the Buy-Quiet initiative, creation of the Safe-in-Sound Award, and noise surveillance.

Literary Views

Roland Barthes distinguishes between physiological noise, which is merely heard, and psychological noise, which is actively listened to. Physiological noise is felt subconsciously as the vibrations of the noise (sound) waves physically interact with the body while psychological noise is perceived as our conscious awareness shifts its attention to that noise.

Luigi Russolo, one of the first composers of noise music, wrote the essay *The Art of Noises*. He argues that any kind of noise could be used as music, as audiences become more familiar with noises caused by technological advancements; noise has become so prominent that pure sound no longer exists.

Henry Cowell claims that technological advancements have reduced unwanted noises from machines, but have not managed so far to completely eliminate them.

References

- Elliot, Barry J. (2002). Designing a Structured Cabling System to ISO 11801 2nd Edition. Cambridge: Woodhead Publishing. p. 80. ISBN 978-0-8247-4130-3.
- Barthes, Roland (1985). The Responsibility of Forms: Critical Essays on Music, Art and Representation. New York: Hill and Wang. ISBN 9780809080755.
- Chilvers, Ian; Glaves-Smith, John, eds. (2009). A Dictionary of Modern and Contemporary Art. Oxford: Oxford University Press. pp. 619–620. ISBN 978-0-19-923965-8.
- Russolo, Luigi (2004). "The art of noises: futurist manifesto". In Cox, Christoph; Warner, Daniel,

eds. Audio Culture: Readings in Modern Music. New York: Continuum. pp. 10ff. ISBN 978-0-8264-1615-5.

- Cowell, Henry (2004). "The joys of noise". In Cox, Christoph; Warner, Daniel, eds. Audio Culture: Readings in Modern Music. New York: Continuum. p. 22. ISBN 978-0-8264-1615-5.
- "EHP – Environmental Noise Pollution in the United States: Developing an Effective Public Health Response". ehp.niehs.nih.gov. Retrieved 2016-06-20.
- Bhatia, Rajiv (May 20, 2014). "Noise Pollution: Managing the Challenge of Urban Sounds". Earth Journalism Network. Retrieved June 23, 2016.
- "Text of S. 3083 (95th): Quiet Communities Act (Passed Congress/Enrolled Bill version) – GovTrack.us". GovTrack.us. Retrieved 2016-06-16.
- "CDC – NIOSH Publications and Products – Criteria for a Recommended Standard: Occupational Exposure to Noise (73-11001)". www.cdc.gov. Retrieved 2016-06-15.
- "Standard Summary Project Fiche: Implementation Capacity for Environmental Noise Directive" (PDF). European Commission. Retrieved June 16, 2016.
- Masterson, Elizabeth (2016-04-27). "Measuring the Impact of Hearing Loss on Quality of Life". Centers for Disease Control and Prevention. Retrieved 2016-06-15.
- "RECOMMENDATION ITU-R BS.468-4 - Measurement of audio-frequency noise voltage" (PDF). www.itu.int. International Telecommunication Union. Retrieved 18 October 2016.
- "Noise-induced Hearing Loss". National Institute on Deafness and Other Communication Disorders (NIDCD). National Institute of Health. March 2014. Retrieved June 16, 2016.
- Murphy, William; Tak, SangWoo (2009-11-24). "Workplace Hearing Loss". Centers for Disease Control and Prevention. Retrieved 2016-06-15.
- Hudson, Heidi; Hayden, Chuck (2011-11-04). "Buy Quiet". Centers for Disease Control and Prevention. Retrieved 2016-06-15.

2

Types of Noise Pollution

This chapter elucidates on certain examples of noise pollution such as noise caused by industries, roadway noise, jet noise, aircraft noise and environmental noise. Noise pollution is as important to deal with as air pollution or water pollution. Understanding noise pollution is necessary if we want to control it in time. Noise pollution is best understood in confluence with the major topics listed in the following chapter.

Line Source

North-South Expressway in Malaysia. A roadway can be a line source of air and noise pollution and need not be a straight line.

A line source, as opposed to a point source, area source, or volume source, is a source of air, noise, water contamination or electromagnetic radiation that emanates from a linear (one-dimensional) geometry. The most prominent linear sources are roadway air pollution, aircraft air emissions, roadway noise, certain types of water pollution sources that emanate over a range of river extent rather than from a discrete point, elongated light tubes, certain dose models in medical physics and electromagnetic antennas. While point sources of pollution were studied since the late nineteenth century, linear sources did not receive much attention from scientists until the late 1960s, when environmental regulations for highways and airports began to emerge. At the same time, computers with the processing power to accommodate the data processing needs of the computer models required to tackle these one-dimensional sources became more available.

In addition, this era of the 1960s saw the first emergence of environmental scientists who spanned the disciplines required to accomplish these studies. For example, meteorologists, chemists, and computer scientists in the air pollution field were required to build complex models to address roadway air dispersion modeling. Prior to the 1960s, these specialities tended to work within their own disciplines, but with the advent of NEPA, the Clean Air Act, the Noise Control Act in the United States, and other seminal legislation, the era of multidisciplinary environmental science had begun.

For electromagnetic linear sources, the principal early advances in computer modeling arose in the Soviet Union and USA when the end of World War II and the Cold War were fought partially by progress in electronic warfare, including the technologies of active antenna arrays.

Linear Air Pollution Source

Air pollution levels near major highways and urban arterials are in violation of U.S. National Ambient Air Quality Standards where millions of Americans live or work. Even the interior of a building does not really protect inhabitants from adverse exterior air quality, since the exterior air is the intake supply, and it is well known that indoor air quality is typically worse than exterior air.

A roadway travelled by motor vehicles can be idealized by a line source emitting air pollutants. This mathematical problem was first solved in 1970 by a collaboration of physics, mathematics and computer science. The original theory assumed steady-state traffic conditions and meteorology on a perfectly straight roadway. Currently the models have evolved to treat variable meteorology, time-variant traffic operations and complex roadbed geometries. Current technology allows highway designers and city planners to analyze alternative roadway development plans and assess air quality impacts. The same basic model theory can be applied to airport operations, since the linear source is merely an inclined line. In the early 1970s these ESL models were refined into area source models to account for the finite width of the roadway.

Linear Noise Source

The New Jersey Turnpike was one of the earliest line sources analyzed for noise

Roadway noise is the most important example of a linear noise source, since it comprises about 80 percent of the environmental noise exposure for humans worldwide. In the 1960s, when computer modeling of this phenomenon was perfected, the first applications of linear source noise modeling became systematic. After passage of the National Environmental Policy Act and Noise Control Act, the demand for detailed analysis soared, and decision makers began to look to acoustical scientists for answers regarding the planning of new roadways and the design of noise mitigation. The intensity of roadway noise is governed by the following variables: traffic operations (speed, truck mix, age of vehicle fleet), roadway surface type, tire types, roadway geometrics, terrain, micrometeorology and the geometry of area structures.

Due to the complexity of the variables, a line source acoustic model must be a computer model that can analyze sound levels in the vicinity of roadways. The first meaningful models arose in the late 1960s and early 1970s. Two of the leading research teams were BBN in Boston and ESL Inc. of Sunnyvale, California. Both of these groups developed complex mathematical models to allow the study of alternate roadway designs, traffic operations and noise mitigation strategies in an arbitrary setting. Later model alterations have come into widespread use among state Departments of Transportation and city planners, but the accuracy of early models has had little change in 40 years.

Generally line source acoustic models trace sound ray bundles and calculate spreading loss along with ray bundle divergence (or convergence} from refractive phenomena. Diffraction is usually addressed by establishing secondary emitters at any points of topographic or anthropomorphic "sharpness" (such as noise barriers or building surfaces. Meteorology can be addressed in a statistical manner allowing for actual wind rose and wind speed statistics (along with thermocline data).

Water Pollution Line Source

Colorado River, receiving effectively a linear source of silt from the sides of the Grand Canyon.

Less common are line source applications in the field of water pollutant dispersal. This phenomenon generally arises when surface runoff scours soil contaminants from upper soil layers and transports these pollutants to a linear receiving water, such as a river.

The underlying land management practices which lead to such sources of water pollution are logging, pesticide application, construction grading, slash-and-burn activity and urban stormwater runoff.

Again computer models are needed to address the complexity of such an extended linear discharge into a dynamic medium such as flowing water. The resulting surface runoff water carrying pollutants may be considered a line source discharging into a river or stream. The chemical composition of this surface runoff may be characterized by a surface runoff model such as the USGS runoff precipitation algorithm, while the instream transport may be analyzed by a dynamic river pollutant model such as DSSAM.

Light Emission Line Source

Common T8 fluorescent lighting tubes used in office environments

In the study of illumination, a variety of sources are linear in nature, most commonly the fluorescent tube, During the process of interior lighting design it is important to calculate the light intensity at work stations or other user areas, not only to ensure sufficient light is present, but more importantly to avoid over-illumination and its attendant energy wastage as well as adverse health effects. Thus the scientists involved in light transmission calculations employ computer models that recognize linear sources when fluorescent fixtures are used. In a typical setting there may be hundreds of finite length light sources that comprise the light output in an office environment. A related concept are the ultraviolet tubes used in phototherapy, where output radiation from the tube can be accurately modeled by treating the tube as a line source. On a larger scale, an illuminated roadway may act as a line source of light pollution.

Industrial Noise

Industrial noise, or occupational noise, is often a term used in relation to environmental health and safety, rather than nuisance, as sustained exposure can cause permanent hearing damage. Industrial noise or occupational noise is the amount of acoustical energy (noise) received by an employees auditory system while they are working.

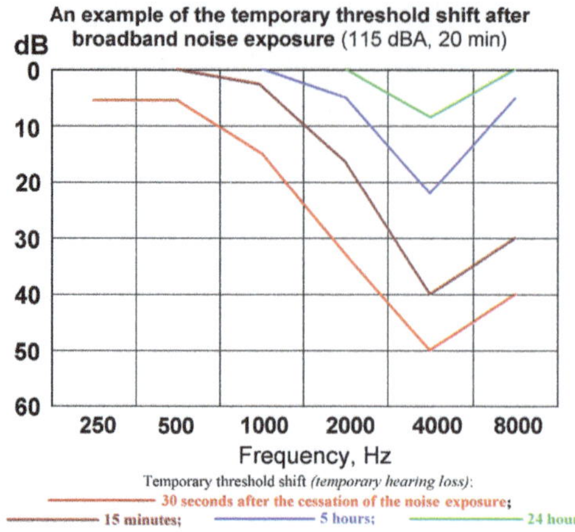

Temporary hearing loss after loud noise exposure. If such exposure is long enough, this temporary threshold shift may become permanent.

“Twenty-two million workers are exposed to potentially damaging noise at work each year. Last year, U.S. business paid more than $1.5 million in penalties for not protecting workers from noise.” - OSHA

Industrial noise is an occupational hazard linked to traditionally loud industries such as ship-building, Mining, railroad work, Welding and Construction. Industrial noise, if experienced repeatedly, at a high intensity, for an extended period of time, can cause noise-induced hearing loss (NIHL). NIHL caused by industrial noise can be classified as occupational hearing loss.

Modern thinking in occupational safety and health further identifies noise as hazardous to workers’ safety and health. This hazard is experienced in various places of employment and through a variety of sources.

Noise, in the context of industrial noise, is hazardous to a persons hearing because of its loud intensity through repeated long-term exposure. In order for Noise to cause Hearing impairment for the worker, the noise has to be close enough, loud enough and the listener has to be exposed for long enough. These factors have been taken into account by the governing occupational health and safety organizations as they determine the unsafe noise exposure levels and durations for their respective industries.

National Institute for Occupational Safety and Health (NIOSH), Occupational Safety and Health Administration (OSHA), Mine Safety and Health Administration (MSHA), Federal Railroad Administration (FRA) have all set standards on hazardous occupational noise in their respective industries. Each industry is different, as workers tasks and equipment differ, but most regulations agree that noise becomes hazardous when it exceeds 85 Decibel, for an 8-hour exposure (typical work shift). This relationship between allotted noise level and exposure time is known as an Exposure action value (EAV) or Permissible exposure limit (PEL). The EAV or PEL can be seen as equations which manipulate the allotted exposure time according to the intensity of the industrial noise. This equation works as an inverse relationship. As the industrial noise intensity increases, the allotted exposure time, to still remain safe, decreases.

These above calculations of PEL and EAV are based on measurements taken to determine the intensity of that particular industrial noise. A-weighted measurements are commonly used to determine noise levels that can cause harm to the human ear. There are also special exposure meters available that integrate noise over a period of time to give an L_{eq} value (equivalent sound pressure level), defined by standards.

Hazardous industrial noise can cause a permanent auditory threshold shift as excessive exposure to loud noises can damage the Hair cells in the ear.

Noise can also effect the safety of the employee and the safety of others. Noise can be a causal factor in work accidents, both by masking hazards and warning signals, and by impeding concentration. High intensity noise can interfere with vital workplace communication which increases the chance of accidents and decreases productivity.

Noise acts synergistically with other hazards to increase the risk of harm to workers. In particular, noise and toxic materials (e.g. some solvents, metals, asphyxiants and pesticides) have some ototoxic properties may also affect the hearing function.

Reduction

Acoustic quieting is the process of making machinery quieter by damping vibrations to prevent them from reaching the observer.

Noise decreases as distance from its source increases. When two identical noise sources are side by side producing a recorded noise of, say, 100 dB(A) the reduction in noise from removing one of the noise sources is about 3 dB, resulting in 97 dB(A). When the distance to a noise source is doubled the recorded noise level is reduced by 6 dB, sometimes called the *Rule of 6*.

The noise attenuation in decibels at a distance from the source d, knowing the SPL at distance d_0, is $20log_{10}\left(\frac{d}{d_0}\right)$. If the distance is doubled, i.e. $\left(\frac{d}{d_0}\right)=2$, the attenuation becomes 6.02 dB (6 for most practical purposes).

Experts have developed a number of standards for noise reduction and isolation of its sources from people.

Regulation

In the United States, the National Institute for Occupational Safety and Health (NIOSH) and the Occupational Safety and Health Administration (OSHA) work together to provide standards and regulations for noise in the workplace. Industrial noise can also be regulated by legislation. A 2012 Cochrane review found low-quality evidence that legislation reduced industrial noise both immediately and in the long-term.

Initiatives

Since the hazards of occupational noise exposure were realized, programs and initiatives such as the US Buy Quiet program have been set up to regulate or discourage noise exposure. The Buy Quiet initiative promotes the purchase of quieter tools and equipment and encourages manufacturers to design quieter equipment. Additionally, the Safe-In-Sound Award was created to recognize successes in hearing loss prevention programs or initiatives.

Roadway Noise

Roadway noise is the most prevalent form of environmental noise. Pictured: São Paulo, Brazil.

Roadway noise is the collective sound energy emanating from motor vehicles. It consists chiefly of road surface, tire, engine/transmission, aerodynamic, and braking elements. In developed and developing countries, roadway noise contributes a proportionately large share of the total societal noise pollution. In the U.S., it contributes more to environmental noise exposure than any other noise source.

History

Roadway noise began to be measured in a widespread manner in the 1960s, when computer modeling of this phenomenon was perfected. After passage of the National Environmental Policy Act and Noise Control Act, the demand for detailed analysis soared, and decision makers began to look to acoustical scientists for answers regarding the planning of new roadways and the design of noise mitigation.

Hong Kong roadway generating noise to adjacent land uses.

Partial bans on motor vehicles from urban areas have been shown to have minimal impacts upon reducing sound levels (as would become clear from later modeling studies); for example, the partial ban in Gothenburg, Sweden resulted in minuscule reduction of sound levels.

Regulation in the EU and Japan of tire and power-train noise has only sought to reduce noise by approx 3 dB, and will only slowly take effect because a few older noisier vehicles can dominate the soundscape.

Traffic operations noise is affected significantly by vehicle speeds, since sound energy roughly doubles for each increment of ten miles an hour in vehicle velocity; an exception to this rule occurs at very low speeds where braking and acceleration noise dominate over aerodynamic noise. Small reductions in vehicle noise occurred in the 1970s as states and provinces enforced unmuffled vehicle ordinances. The vehicle fleet noise has not changed very much over the last three decades; however, if the trend in hybrid vehicle use continues, substantial noise reduction will occur, especially in the regime of traffic flow below 35 miles per hour. Hybrid vehicles are so quiet at low speeds that they create a pedestrian safety issue when reversing or maneuvering when parking etc. (but not when going travelling forward), and so are typically fitted with electric vehicle warning sounds.

Trucks contribute a disproportionate amount of noise not only because of their large engines, but also the height of the diesel stack and the aerodynamic drag. Significant interior noise is usually present inside moving motor vehicles; in fact, passengers are generally not aware that these levels are high, because experience has led motorists to expect levels commonly exceeding 65 dBA.

Roadway surface types contribute to different noise levels. Of the common types of surfaces in modern cities, there is a 4 dB difference between the loudest and the softest: chip seal type and grooved roads being the loudest, and concrete surfaces without spacers being the quietest, and asphaltic surfaces being about average. Rubberized asphalt (which uses recycled old tyres) is much quieter and is already widely used. Experimental poro-elastic road surfacing might halve road noise.

Tire types can cause 10 dB(A) variations in noise, based on a 2001 sample of 100 commercially available tires. As of 2001, there was no correlation between grip and noise. Quieter tires may have slightly lower rolling resistance. Tire labeling for noise, grip, and rolling resistance has been widely introduced in Europe, with noisy tires being taxed.

Roadway geometrics and surrounding terrain are interrelated, since the propagation of sound is sensitive to the overall geometry and must consider diffraction (bending of sound waves around obstacles), reflection, ground wave attenuation, spreading loss and refraction. A simple discussion indicates that sound will be diminished when the path of sound is blocked by terrain, or will be enhanced if the roadway is elevated so as to broadcast; however, the complexities of variable interaction are so great, that there are many exceptions to this simple argument.

Micrometeorology is significant in that sound waves can be refracted by wind gradients or thermoclines, effectively dismissing the effect of some noise barriers or terrain intervention.

Geometry of area structures is an important input, since the presence of buildings or walls can block sound under certain circumstances, but reflective properties can augment sound energy at other locations.

Computer Models

At a macro level, ongoing research is required for national and world-wide responses to road noise pollution - issues include road surface choices, the regulation and taxing of noisy designs, and the on-going inspecting of individual vehicles.

At the micro level of managing particular roads, because of the complexity of the variables discussed above, it is necessary to create a computer model that can analyze sound levels in the vicinity of roadways. The first meaningful models arose in the late 1960s and early 1970s addressing the noise line source (e.g. roadway). Two of the leading research teams were BBN in Boston and ESL of Sunnyvale, California. Both of these groups developed complex mathematical models to allow the study of alternate roadway designs, traffic operations and noise mitigation strategies in an arbitrary setting. Later model alterations have come into widespread use among state Departments of

Transportation and city planners, but the accuracy of early models has had little change in 40 years.

Generally the models trace sound ray bundles and calculate spreading loss along with ray bundle divergence (or convergence) from refractive phenomena. Diffraction is usually addressed by establishing secondary emitters at any points of topographic or anthropomorphic "sharpness" (such as noise barriers or building surfaces). Meteorology can be addressed in a statistical manner allowing for actual wind rose and wind speed statistics (along with thermocline data). Recent models have also attempted to predict levels of local air pollution based on an analysis of specific frequencies that are related to tire and engine noise.

Case Studies of Road Design

An interesting early case where two of the leading models were pitted against each other involved a proposed widening of the New Jersey Turnpike from six to twelve lanes. The BBN and ESL models were on opposing sides of a matter decided in New Jersey Superior Court. This case in the early 1970s was one of the first U.S. examples of acoustical scientists playing a role in the design of a major highway. The models allowed the court to understand the effects of roadway geometry (width in this case), vehicle speeds, proposed noise barriers, residential setback and pavement types. The outcome was a compromise that involved substantial mitigation of noise pollution impacts.

Another early case involved the proposed extension of Interstate 66 through Arlington, Virginia. The plaintiff, Arlington Coalition on Transportation sued the Virginia Department of Transportation on the grounds of air quality, noise and neighborhood disruption. To analyze roadway noise, the ESL model was used by the plaintiff, who won this case partially due to the credibility of the computer model. The matter was revisited a decade later and a greatly reduced highway design with transit element and extensive noise mitigation was agreed to.

Later cases have occurred in every state, both in contentious actions and in routine highway planning and design. The public as well as governmental agencies have become aware of the value of acoustical science to provide useful insights to the roadway design process.

Worldwide Perspective

Even without regulation, there are strong individual economic pressures for quieter vehicles, because owners and employers see quieter vehicles as more luxurious and less stressful. The tighter regulatory requirements of the EU and Japan encourage quieter design even in unregulated countries, because most car manufacturers aspire to international sales. On the other hand, individual owners of motorbikes, 'boom-box' cars (with very loud music systems), and 'muscle-cars' may prefer their vehicle to be louder (at least at idling or low speeds), and such noise (often from modified exhaust systems) can only be controlled by on-going inspection and sanctions.

Several studies have concluded that reducing traffic noise pollution is low-cost or cost-effective. Such studies include consideration of the reduced value of noise-affected real-estate, the costs of supporting a dispersed population 'trying to get away from all the noise', and the increased healthcare costs statistically attributable to a noisier environment.

European technology began to emulate the United States treatment of roadway noise by the 1980s, although the national requirements of noise studies generally remain less stringent than the U.S. In developing countries, noise pollution from motor vehicles represents a significant impact, but technologies are not as advanced as in Western nations. For example, a recent paper from Iran illustrates a level of technology that the United States encountered in the 1960s. The European Union has recently proposed a set of vehicle tire requirements, similar to those introduced in the U.S. in the 1970s.

Aircraft Noise

Aircraft noise is noise pollution produced by any aircraft or its components, during various phases of a flight: on the ground while parked such as auxiliary power units, while taxiing, on run-up from propeller and jet exhaust, during take off, underneath and lateral to departure and arrival paths, over-flying while en route, or during landing.

Mechanisms of Sound Production

Small general aviation aircraft produce localized aircraft noise.

Helicopter main and tail rotors produce aerodynamic noise.

A moving aircraft including the jet engine or propeller causes compression and rarefaction of the air, producing motion of air molecules. This movement propagates through the air as pressure waves. If these pressure waves are strong enough and within the audible frequency spectrum, a sensation of hearing is produced. Different aircraft types have different noise levels and frequencies. The noise originates from three main sources:

- Aerodynamic noise
- Engine and other mechanical noise
- Noise from aircraft systems

Aerodynamic Noise

Aerodynamic noise arises from the airflow around the aircraft fuselage and control surfaces. This type of noise increases with aircraft speed and also at low altitudes due to the density of the air. Jet-powered aircraft create intense noise from aerodynamics. Low-flying, high-speed military aircraft produce especially loud aerodynamic noise.

The shape of the nose, windshield or canopy of an aircraft affects the sound produced. Much of the noise of a propeller aircraft is of aerodynamic origin due to the flow of air around the blades. The helicopter main and tail rotors also give rise to aerodynamic noise. This type of aerodynamic noise is mostly low frequency determined by the rotor speed.

Typically noise is generated when flow passes an object on the aircraft, for example the wings or landing gear. There are broadly two main types of airframe noise:

- Bluff Body Noise – the alternating vortex shedding from either side of a bluff body, creates low pressure regions (at the core of the shed vortices) which manifest themselves as pressure waves (or sound). The separated flow around the bluff body is quite unstable, and the flow "rolls up" into ring vortices—which later break down into turbulence.

- Edge Noise – when turbulent flow passes the end of an object, or gaps in a structure (high lift device clearance gaps) the associated fluctuations in pressure are heard as the sound propagates from the edge of the object (radially downwards).

Engine and Other Mechanical Noise

Much of the noise in propeller aircraft comes equally from the propellers and aerodynamics. Helicopter noise is aerodynamically induced noise from the main and tail rotors and mechanically induced noise from the main gearbox and various transmission chains. The mechanical sources produce narrow band high intensity peaks relating to the rotational speed and movement of the moving parts. In computer modelling terms noise from a moving aircraft can be treated as a line source.

Aircraft gas turbine engines (jet engines) are responsible for much of the aircraft noise during takeoff and climb, such as the *buzzsaw noise* generated when the tips of the fan blades reach supersonic speeds. However, with advances in noise reduction technologies—the airframe is typically more noisy during landing.

The majority of engine noise is due to jet noise—although high bypass-ratio turbofans do have considerable fan noise. The high velocity jet leaving the back of the engine has an inherent shear layer instability (if not thick enough) and rolls up into ring vortices. This later breaks down into turbulence. The SPL associated with engine noise is proportional to the jet speed (to a high power) therefore, even modest reductions in exhaust velocity will produce a large reduction in Jet Noise.

Noise from Aircraft Systems

Cockpit and cabin pressurization and conditioning systems are often a major contributor within cabins of both civilian and military aircraft. However, one of the most significant sources of cabin noise from commercial jet aircraft, other than the engines, is the Auxiliary Power Unit (APU), an onboard generator used in aircraft to start the main engines, usually with compressed air, and to provide electrical power while the aircraft is on the ground. Other internal aircraft systems can also contribute, such as specialized electronic equipment in some military aircraft.

Health Effects

There are health consequences of elevated sound levels. Elevated workplace or other noise can cause hearing impairment, hypertension, ischemic heart disease, annoyance, sleep disturbance, and decreased school performance. Although some hearing loss occurs naturally with age, in many developed nations the impact of noise is sufficient to impair hearing over the course of a lifetime. Elevated noise levels can create stress, increase workplace accident rates, and stimulate aggression and other anti-social behaviors.

German Environmental Study

A large-scale statistical analysis of the health effects of aircraft noise was undertaken in the late 2000s by Bernhard Greiser for the Umweltbundesamt, Germany's central environmental office. The health data of over one million residents around the Cologne airport were analysed for health effects correlating with aircraft noise. The results were then corrected for other noise influences in the residential areas, and for socioeconomic factors, to reduce possible skewing of the data.

The German study concluded that aircraft noise clearly and significantly impairs health. For example, a day-time average sound pressure level of 60 decibel increasing coronary heart disease by 61% in men and 80% in women. As another indicator, a night-time average sound pressure level of 55 decibel increased the risk of heart attacks by 66% in men and 139% in women. Statistically significant health effects did however start as early as from an average sound pressure level of 40 decibel.

FAA Advice

The FAA says that a maximum day-night average sound level of 65 dB is incompatible with residential communities. Communities in affected areas may eligible for mitigation such as soundproofing.

Inside Aircraft

Noise associated with aircraft does not only affect people on the ground, but also those within the aircraft (e.g., flight crew, cabin crew and passengers). While there appears to be little research in this area, lower noise inside the aircraft is widely promoted as a benefit for new aircraft. The noise levels inside an Airbus A321 during cruise have been reported as approximately 78 dB(A). During taxi when the aircraft engines are producing minimal thrust, noise levels in the cabin have been recorded at 65 dB(A). This is approximately 20 decibels louder than recommended acceptable levels for an office but 20 decibels below the occupational noise exposure limits of 85 dB(A).

Cognitive Effects

Simulated aircraft noise at 65 dB(A) has been shown to negatively affect individuals' memory and recall of auditory information. In one study comparing the effect of aircraft noise to the effect of alcohol on cognitive performance, it was found that simulated aircraft noise at 65 dB(A) had the same effect on individuals' ability to recall auditory information as being intoxicated with a Blood Alcohol Concentration (BAC) level of at 0.10. A BAC of 0.10 is double the legal limit required to operate a motor vehicle in many developed countries such as Australia.

Noise Mitigation Programs

Noise-reducing chevrons on a Rolls-Royce Trent 1000 turbofan engine.

In the United States, since aviation noise became a public issue in the late 1960s, governments have enacted legislative controls. Aircraft designers, manufacturers, and operators have developed quieter aircraft and better operating procedures. Modern high-bypass turbofan engines, for example, are quieter than the turbojets and low-bypass turbofans of the 1960s. First, FAA Aircraft Certification achieved noise reductions classified as "Stage 3" aircraft; which has been upgraded to "Stage 4" noise certification resulting in quieter aircraft. This has resulted in lower noise exposures in spite of increased traffic growth and popularity.

In the 1980s the U.S. Congress authorized the FAA to devise programs to insulate homes near airports. While this does not address the external noise, the program has been effective for residential interiors. Some of the first airports at which the technology was applied were San Francisco International Airport and San Jose International Airport in California. A computer model is used which simulates the effects of aircraft noise upon building structures. Variations of aircraft type, flight patterns and local meteorology can be studied. Then the benefits of building retrofit strategies such as roof upgrading, window glazing improvement, fireplace baffling, caulking construction seams can be evaluated.

Night Flying Restrictions

At Heathrow, Gatwick and Stansted airports in London, UK and Frankfurt Airport in Germany, night flying restrictions apply to reduce noise exposure at night.

Satellite-based Navigation Systems

A series of trials were undertaken at London's Heathrow Airport, between December 2013 and November 2014, as part of the UK's "Future Airspace Strategy", and the Europe-wide "Single European Sky" modernisation project. The trials demonstrated that using satellite-based navigation systems it was possible to offer noise relief to more surrounding communities. The study found that steeper angles for take-off and landing led to fewer people experiencing aircraft noise and that noise relief could be shared by using more precise flight paths, allowing control of the noise footprint of departing aircraft. Noise relief could be enhanced by switching flight paths, for example by using one flight path in the morning and another in the afternoon.

Jet Noise

In aeroacoustics, jet noise is the field that focuses on the noise generation caused by high-velocity jets and the turbulent eddies generated by shearing flow. Such noise is known as broadband noise and extends well beyond the range of human hearing (100 kHz and higher). Jet noise is also responsible for some of the loudest sounds ever produced by mankind.

Sources of Jet Noise

The primary sources of jet noise for a high-speed air jet (meaning when the exhaust velocity exceeds about 100 m/s) are "jet mixing noise" and, for supersonic flow, shock associated noise. Also, acoustic sources within the "jet pipe" also contribute to the noise, mainly at lower speeds, which include combustion noise and sounds produced by interactions of a turbulent stream with fans, compressors, and turbine systems.

The jet mixing sound is created by the turbulent mixing of a jet with the ambient fluid, in most cases, air. The mixing initially occurs in an annular shear layer, which grows with the length of the nozzle. The mixing region generally fills the entire jet at four or five diameters from the nozzle. The high-frequency components of the sound are mainly stationed close to the nozzle, where the dimensions of the turbulence eddies are small. Further down the jet, where the eddy size is similar to the jet diameter, is where lower frequency begins.

In Supersonic, or "choked" jets there are cells through which the flow continuously expands and contracts. Several of these "shock cells" can be seen extending up to ten jet diameters from the nozzle and are responsible for two additional components of jet noise, "screech tones" and broadband "shock associated noises". Screech is produced by a feedback mechanism in which a disturbance convecting in the shear layer generates sound as it traverses the standing system of shock waves in the jet. Even though screech is a side effect of the jet's flight, it can be suppressed by an appropriate design for a nozzle.

Aircraft noise is also sometimes called "jet noise" when emanating from jet aircraft, regardless of the mechanism of noise production.

Environmental Noise

Environmental noise is the summary of noise pollution from outside, caused by transport, industrial and recreational activities.

Noise is frequently described as 'unwanted sound', and, within this context, environmental noise is generally present in some form in all areas of human activity. The effects in humans of exposure to environmental noise may vary from emotional to physiological and psychological.

Noise at low levels is not necessarily harmful; environmental noise can also convey a sense of liveliness in an area, and is not then always considered 'unwanted'. However, the adverse effects of noise exposure (i.e. noise pollution) could include: interference with speech or other 'desired' sounds, annoyance, sleep disturbance, anxiety, hearing damage and stress-related cardiovascular health problems.

As a result, environmental noise is studied, regulated and monitored by many governments and institutions, as well as forming the basis or branch for a number of different occupations.

Environmental Noise Emission

Noise from transportation is typically emitted by the machinery (e.g. the engine or exhaust) and aerodynamic noise caused by the compression and friction in the air around the vessel during motion.

Industrial and recreational noise could be generated by a multitude of different sources and processes.

Sound propagation outdoors is subject to meteorological effects (e.g. wind, temperature) that affect the distance, speed, and direction with which environmental noise travels from a source to a listener.

Environmental Noise Policy and Regulation

European Union

The European Union has a special definition based on the European directive 2002/49/EC. This directive gives a definition for environmental noise. The main target is an integrated noise management.

The implementation is divided into phases: In the first phase the member states shall inform about major roads with more than six million vehicles a year, major railways with have more than 60,000 trains per year, major airports with more than 50,000 movements per year and agglomerations with more than 250,000 inhabitants. The second phase started with lower, partly the halve numbers. Only the criteria for airports remained unchanged. In the third and the following phases the methods for calculation the noise levels will change but the describd criterias remain unchanged. Each phase consists of three steps: the collection of the data about the main sources of noise as described before, strategic noise maps and action plans.

Austria

In Austria the institution which is responsible for the noise sources is also responsible for the noise maps concerning these sources. This means that the Federation is responsible for the federal roads and each state is responsible for the country's roads.

France

France reported 24 agglomeration areas, Paris was the biggest with 9.6 million inhabitants and 272 square kilometres.

Germany

In Germany recreational activities are not considered to be part of environmental noise. Germany has implemented national regulations in 2005 and 2006 and reported 27 agglomeration areas in the first phase: Berlin was the biggest with 3.39 million inhabitants and 889 square kilometres, Hamburg the largest with 1,045 square kilometres and 2 million inhabitants. The smallest was Gelsenkirchen with 270,000 inhabitants and 105 square kilometres. In the national legislation noise resulting from recreational activities like sports and leisure is not considered as environmental noise.

United Kingdom

The United Kingdom reported a total of 28 metropolitan areas, where London is the largest with 8.3 million inhabitants. The majority of metropolitan areas are located in England; in Scotland and Wales there are each two, in Northern Ireland only the capital Belfast.

White Noise

Colors of noise
White
Pink
Red (Brownian)
Grey

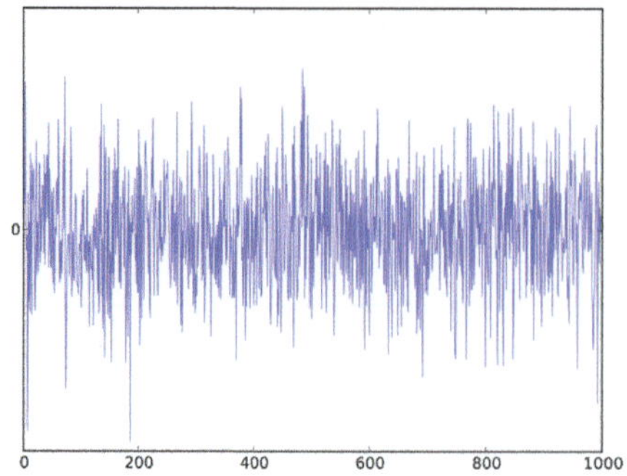

Plot of a Gaussian white noise signal

In signal processing, white noise is a random signal having equal intensity at different frequencies, giving it a constant power spectral density. The term is used, with this or similar meanings, in many scientific and technical disciplines, including physics, acoustic engineering, telecommunications, statistical forecasting, and many more. White noise refers to a statistical model for signals and signal sources, rather than to any specific signal.

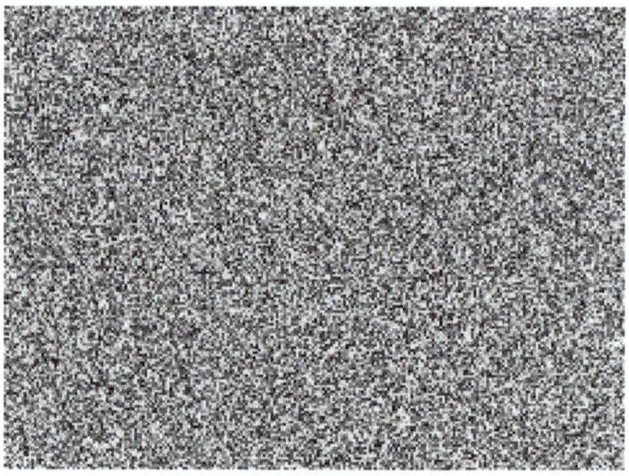

A "white noise" image

In discrete time, white noise is a discrete signal whose samples are regarded as a sequence of serially uncorrelated random variables with zero mean and finite variance; a single realization of white noise is a random shock. Depending on the context, one may also require that the samples be independent and have identical probability distribution (in other words i.i.d. is a simplest representative of the white noise). In particular, if each sample has a normal distribution with zero mean, the signal is said to be Gaussian white noise.

The samples of a white noise signal may be sequential in time, or arranged along one or more spatial dimensions. In digital image processing, the pixels of a *white noise image* are typically arranged in a rectangular grid, and are assumed to be independent random variables with uniform probability distribution over some interval. The concept can be defined also for signals spread over more complicated domains, such as a sphere or a torus.

An infinite-bandwidth white noise signal is a purely theoretical construction. The bandwidth of white noise is limited in practice by the mechanism of noise generation, by the transmission medium and by finite observation capabilities. Thus, a random signal is considered "white noise" if it is observed to have a flat spectrum over the range of frequencies that is relevant to the context. For an audio signal, for example, the relevant range is the band of audible sound frequencies, between 20 and 20,000 Hz. Such a signal is heard as a hissing sound, resembling the /sh/ sound in "ash". In music and acoustics, the term "white noise" may be used for any signal that has a similar hissing sound.

White noise draws its name from white light, although light that appears white generally does not have a flat spectral power density over the visible band.

The term white noise is sometimes used in the context of phylogenetically based statistical methods to refer to a lack of phylogenetic pattern in comparative data. It is sometimes used in non technical contexts, in the metaphoric sense of "random talk without meaningful contents".

Statistical Properties

Spectrogram of pink noise (left) and white noise (right), shown with linear frequency axis (vertical)

Being uncorrelated in time does not restrict the values a signal can take. Any distribution of values is possible (although it must have zero DC component). Even a binary signal which can only take on the values 1 or –1 will be white if the sequence is statistically uncorrelated. Noise having a continuous distribution, such as a normal distribution, can of course be white.

It is often incorrectly assumed that Gaussian noise (i.e., noise with a Gaussian amplitude distribution) necessarily refers to white noise, yet neither property implies the other. Gaussianity refers to the probability distribution with respect to the value, in this context the probability of the signal falling within any particular range of amplitudes, while the term 'white' refers to the way the signal power is distributed (i.e., independently) over time or among frequencies.

We can therefore find Gaussian white noise, but also Poisson, Cauchy, etc. white noises. Thus, the two words "Gaussian" and "white" are often both specified in mathematical models of systems. Gaussian white noise is a good approximation of many real-world situations and generates mathematically tractable models. These models are used so frequently that the term additive white Gaussian noise has a standard abbreviation: AWGN.

White noise is the generalized mean-square derivative of the Wiener process or Brownian motion.

A generalization to random elements on infinite dimensional spaces, such as random fields, is the white noise measure.

Practical Applications

Music

White noise is commonly used in the production of electronic music, usually either directly or as an input for a filter to create other types of noise signal. It is used extensively in audio synthesis, typically to recreate percussive instruments such as cymbals or snare drums which have high noise content in their frequency domain.

Electronics Engineering

White noise is also used to obtain the impulse response of an electrical circuit, in particular of amplifiers and other audio equipment. It is not used for testing loudspeakers as its spectrum contains too great an amount of high frequency content. Pink noise, which differs from white noise in that it has equal energy in each octave, is used for testing transducers such as loudspeakers and microphones.

Acoustics

To set up the equalization for a concert or other performance in a venue, a short burst of white or pink noise is sent through the PA system and monitored from various points in the venue so that the engineer can tell if the acoustics of the building naturally boost or cut any frequencies. The engineer can then adjust the overall equalization to ensure a balanced mix.

Computing

White noise is used as the basis of some random number generators. For example, Random.org uses a system of atmospheric antennae to generate random digit patterns from white noise.

Tinnitus Treatment

White noise is a common synthetic noise source used for sound masking by a tinnitus masker. White noise machines and other white noise sources are sold as privacy enhancers and sleep aids and to mask tinnitus. Alternatively, the use of an FM radio tuned to unused frequencies ("static") is a simpler and more cost-effective source of white noise. However, white noise generated from a common commercial radio receiver tuned to an unused frequency is extremely vulnerable to being contaminated with spurious signals, such as adjacent radio stations, harmonics from non-adjacent radio stations, electrical equipment in the vicinity of the receiving antenna causing interference, or even atmospheric events such as solar flares and especially lightning.

Work Environment

The effects of white noise upon cognitive function are mixed. Recently, a small study found that white noise background stimulation improves cognitive functioning among secondary students with attention deficit hyperactivity disorder (ADHD), while decreasing performance of non-ADHD students. Other work indicates it is effective in improving the mood and performance of workers by masking background office noise, but decreases cognitive performance in complex card sorting tasks.

Mathematical Definitions

White Noise Vector

A random vector (that is, a partially indeterminate process that produces vectors of real numbers) is said to be a white noise vector or white random vector if its components each have a probability distribution with zero mean and finite variance, and are statistically uncorrelated: that is, their joint probability distribution must be the product of the distributions of the individual components.

A necessary (but, in general, not sufficient) condition for statistical independence of two variables is that they be statistically uncorrelated; that is, their covariance is zero. Therefore, the covariance matrix R of the components of a white noise vector w with n elements must be an n by n diagonal matrix, where each diagonal element R_{ii} is the variance of component w_i; and the correlation matrix must be the n by n identity matrix.

In particular, if in addition to being independent every variable in w also has a normal distribution with zero mean and the same variance σ^2, w is said to be a **Gaussian** white noise vector. In that case, the joint distribution of w is a multivariate normal distribution; the independence between the variables then implies that the distribution has spherical symmetry in n-dimensional space. Therefore, any orthogonal transformation of the vector will result in a Gaussian white random vector. In particular, under most types of discrete Fourier transform, such as FFT and Hartley, the transform W of w will be a Gaussian white noise vector, too; that is, the n Fourier coefficients of w will be independent Gaussian variables with zero mean and the same variance σ^2.

The power spectrum P of a random vector w can be defined as the expected value of the squared modulus of each coefficient of its Fourier transform W, that is, $P_i = E(|W_i|^2)$. Under that definition, a Gaussian white noise vector will have a perfectly flat power spectrum, with $P_i = \sigma^2$ for all i.

If w is a white random vector, but not a Gaussian one, its Fourier coefficients W_i will not be completely independent of each other; although for large n and common probability distributions the dependencies are very subtle, and their pairwise correlations can be assumed to be zero.

Often the weaker condition "statistically uncorrelated" is used in the definition of white noise, instead of "statistically independent". However some of the commonly expected properties of white noise (such as flat power spectrum) may not hold for this weaker version. Under this assumption, the stricter version can be referred to explicitly as independent white noise vector. Other authors use strongly white and weakly white instead.

An example of a random vector that is "Gaussian white noise" in the weak but not in the strong sense is $x=[x_1,x_2]$ where x_1 is a normal random variable with zero mean, and x_2 is equal to $+x_1$ or to $-x_1$, with equal probability. These two variables are uncorrelated and individually normally distributed, but they are not jointly normally distributed and are not independent. If x is rotated by 45 degrees, its two components will still be uncorrelated, but their distribution will no longer be normal.

In some situations one may relax the definition by allowing each component of a white random vector w to have non-zero expected value μ. In image processing especially, where samples are typically restricted to positive values, one often takes μ to be one half of the maximum sample value. In that case, the Fourier coefficient W_0 corresponding to the zero-frequency component (essentially, the average of the w_i) will also have a non-zero expected value $\mu\sqrt{n}$; and the power spectrum P will be flat only over the non-zero frequencies.

Continuous-time White Noise

In order to define the notion of "white noise" in the theory of continuous-time signals, one must replace the concept of a "random vector" by a continuous-time random signal; that is, a random process that generates a function w of a real-valued parameter t.

Such a process is said to be white noise in the strongest sense if the value $w(t)$ for any time is a random variable that is statistically independent of its entire history before t. A weaker definition requires independence only between the values t and $w(t_1)$ at every pair of distinct times t_1 and t_2. An even weaker definition requires only that such pairs $w(t_1)$ and $w(t_2)$ be uncorrelated. As in the discrete case, some authors adopt the weaker definition for "white noise", and use the qualifier independent to refer to either of the stronger definitions. Others use weakly white and strongly white to distinguish between them.

However, a precise definition of these concepts is not trivial, because some quantities that are finite sums in the finite discrete case must be replaced by integrals that may not converge. Indeed, the set of all possible instances of a signal w is no longer a finite-dimensional space , but an infinite-dimensional function space. Moreover, by any definition a white noise signal $\mathbb{R}^n$ would have to be essentially discontinuous at every point; therefore even the simplest operations on w, like integration over a finite interval, require advanced mathematical machinery.

Some authors require each value w_t to be a real-valued random variable with some finite variance $w(t)$. Then the covariance σ^2. between the values at two times $\mathrm{E}(w(t_1)\cdot w(t_2))$ and t_1 is well-defined: it is zero if the times are distinct, and t_2 if they are equal. However, by this definition, the integral

$$W_{[a,a+r]} = \int_a^{a+r} w(t)dt$$

over any interval with positive width r would be zero. This property would render the concept inadequate as a model of physical "white noise" signals.

Therefore, most authors define the signal w indirectly by specifying non-zero values for the integrals of $w(t)$ and $|w(t)|^2$ over any interval $[a, a+r]$, as a function of its width r. In this approach, however, the value of $w(t)$ at an isolated time cannot be defined as a real-valued random variable. Also the covariance $\mathrm{E}(w(t_1)\cdot w(t_2))$ becomes infinite when $t_1 = t_2$; and the autocorrelation function $\mathrm{R}(t_1, t_2)$ must be defined as $N\delta(t_1 - t_2)$,, where N is some real constant and δ is Dirac's "function".

In this approach, one usually specifies that the integral W_I of $w(t)$ over an interval $I = [a,b]$ is a real random variable with normal distribution, zero mean, and variance $(b-a)\sigma^2$; and also that the covariance $\mathrm{E}(W_I \cdot W_J)$ of the integrals is where is the width of the intersection $I \cap J$ of the two intervals I, J. This model is called a Gaussian white noise signal (or process).

Mathematical Applications

Time Series Analysis and Regression

In statistics and econometrics one often assumes that an observed series of data values is the sum of a series of values generated by a deterministic linear process, depending on certain independent (explanatory) variables, and on a series of random noise values. Then regression analysis is used to infer the parameters of the model process from the observed data, e.g. by ordinary least squares, and to test the null hypothesis that each of the parameters is zero against the alternative hypothesis that it is non-zero. Hypothesis testing typically assumes that the noise values are mutually uncorrelated with zero mean and the same Gaussian probability distribution – in other words, that the noise is white. If there is non-zero correlation between the noise values underlying different observations then the estimated model parameters are still unbiased, but estimates of their uncertainties (such as confidence intervals) will be biased (not accurate on average). This is also true if the noise is heteroskedastic – that is, if it has different variances for different data points.

Alternatively, in the subset of regression analysis known as time series analysis there are often no explanatory variables other than the past values of the variable being modeled (the dependent variable). In this case the noise process is often modeled as a moving average process, in which the current value of the dependent variable depends on current and past values of a sequential white noise process.

Random Vector Transformations

These two ideas are crucial in applications such as channel estimation and channel equalization in communications and audio. These concepts are also used in data compression.

In particular, by a suitable linear transformation (a coloring transformation), a white random vector can be used to produce a "non-white" random vector (that is, a list of random variables) whose elements have a prescribed covariance matrix. Conversely, a random vector with known covariance matrix can be transformed into a white random vector by a suitable whitening transformation.

Generation

White noise may be generated digitally with a digital signal processor, microprocessor, or microcontroller. Generating white noise typically entails feeding an appropriate stream of random numbers to a digital-to-analog converter. The quality of the white noise will depend on the quality of the algorithm used.

References

- Izmerov, Nikolai; Suvorov, Herman; Prokopenko, Ludmila (2001). "Chapter 4. Occupational hearing loss". The man and the noise (Человек и шум) (in Russian). Moscow: ГЕОТАР-МЕД. p. 103. ISBN 5-9231-0057-6.
- Kryter, Karl D. (1994). The handbook of hearing and the effects of noise: physiology, psychology, and public health. Boston: Academic Press. ISBN 0-12-427455-2.
- Hogan, C. Michael and Jorgen Ravnkilde, Design of acoustical insulation for existing residences in the vicinity of San Jose Municipal Airport, 1 January 1984, FAA grant-funded research, ISBN B0007B2OG0
- Howe, M.S. (1998). Acoustics of Fluid-Structure Interactions. Cambridge: Cambridge University Press. pp. 149–156. ISBN 978-0-521-63320-8.
- "Safety and Health Topics | Occupational Noise Exposure | Occupational Safety and Health Administration". www.osha.gov. Retrieved 2016-11-08.
- Johnson and Morata, TC (2010). "Occupational exposure to chemicals and hearing impairment" (PDF). Arbete och Hälsa. 44 (4): 177 pp. Retrieved 27 June 2016.

- "Occupational Noise Exposure: Standards". Occupational Safety and Health Administration. Retrieved 2016-07-14.
- "Noise and Hearing Loss Prevention: National Goals, Policies, and Standards". Centers for Disease Control and Prevention. Retrieved 2016-07-14.
- Sandberg, Ulf. "Tyre/road noise – Myths and realities" (PDF). The 2001 International Congress and Exhibition on Noise Control Engineering The Hague, The Netherlands, 2001 August 27–30. Retrieved 2013-07-25.

3

Sound Level Measurements

To mark a difference between sound and noise we need some level of measurement. Sound exposure level and sound intensity level are two levels explained within this chapter. The following section helps in providing the reader with an in-depth understanding of the subject matter.

Noise Calculation

noise calculation is the process of calculating the level of noise immission using the metric dB(A). Noise immission is created by noise sources (noise emission) of various types which are propagating noise into the environment. A single source will create a certain level of immission primarily driven by originating sound power level and distance influenced by e.g. absorption and reflection. Several noise sources result in typically higher levels of immission. The method or process of determining the resulting immission level is called noise calculation, its graphical representation is called noise map.

Noise calculation is done for the following reasons

* to calculate a current situation based on existing physics, mostly when a physical measurement at a location is impractical

* to predict the resulting immission levels based on a planned change, e.g. set up of a new machine

Mostly, noise calculation is part of any such planning process and may become part of the decision process for physical change

Environment engineering differentiates between

* indoor sound propagation, e.g. occupational health considerations

* outdoor sound propagation, e.g. city planning considerations

as well as type of noise source, e.g.

* industrial noise

* rail or air traffic noise

* road noise

Each of before mentioned require different algorithms for noise calculation which have been defined over time in a number of different national or international standards. As an example, outdoor sound propagation is defined in ISO standard 9613 which can be received from the ISO organisation or scientific literature.

Calculation Standards

Following standards for noise calculation have been agreed in the past:

'*ISO 9613 parts 1 and 2*' defines the calculation methods for outdoor sound propagation (International)

'*VDI2714 and VDI2720* ' Schallausbreitung und Schallschutz im Freien (outdoor sound propagation, Germany)

'*DIN18005* ' Schallschutz im Städtebau (noise protection in cities, Germany)

'*TA Laerm* ' Technische Anleitung zum Schutz gegen Lärm (Germany)

'*VDI-Richtlinie 3745 Blatt 1* ' Beurteilung von Schießgeräuschimmissionen (Germany)

'*BS 5228* ' Code of practice for noise and vibration control on construction and open sites (UK)

'*RLS-90* ' Richtlinien für den Lärmschutz an Straßen (noise protection at roads/traffic, Germany)

'*CRTN* ' calculation of road traffic noise (UK)

'*Schall03* ' Schienenlärm (calculation of rail noise, Germany)

Calculation Tools

There is limited number of calculation tools that have been developed primarily by noise experts and small specialized companies. Those have grown over time becoming more user friendly, covering more application cases and adding service elements to the noise calculation tools. Since the process of noise calculation is complex, both from an input perspective (gathering data, modelling the acoustic elements in the field correctly) and from a pure compute power perspective more recently advanced trends wer arising

* moving the calculation routines into self-contained calculation engines
* adapting CAD systems for input of acoustic models
* moving into 3D for displaying models or even data capture
* moving into cloud services and web enabled calculation tools.

Afterall, the quality of results from the noise calculation very much depends on the quality of the acoustic data provided and the capability of the noise calculation engine.o take decisions.

Some of the software packages more used for noise calculation are:

- noise3D online
- Code_Tympan
- IMMI
- Olive Tree Lab Terrain

Some online tools used for noise calculations are:

- acoustics.io - construction noise calculator

Sound Exposure

Sound exposure is the integral, over time, of squared sound pressure. The SI unit of sound exposure is the pascal squared second (Pa²·s).

Mathematical Definition

Sound exposure, denoted *E*, is defined by

$$E = \int_{t_0}^{t_1} p(t)^2 \, \mathrm{d}t,$$

where

- the exposure is being calculated for the time interval between times t_0 and t_1;
- *p(t)* is the sound pressure at time *t*, usually A-weighted for sound in air.

Sound Exposure Level

Sound exposure level (SEL) or acoustic exposure level is a logarithmic measure of the sound exposure of a sound relative to a reference value. Sound exposure level, denoted L_E and measured in dB, is defined by

$$L_E = \frac{1}{2}\ln\left(\frac{E}{E_0}\right)\mathrm{Np} = \log_{10}\left(\frac{E}{E_0}\right)\mathrm{B} = 10\log_{10}\left(\frac{E}{E_0}\right)\mathrm{dB},$$

where

- E is the sound exposure;
- E_0 is the reference sound exposure;
- 1 Np = 1 is the neper;
- 1 B = 1/2 ln 10 is the bel;
- 1 dB = 1/20 ln 10 is the decibel.

The commonly used reference sound exposure in air is

$$E_0 = 400\,\mu\text{Pa}^2 \cdot \text{s}.$$

The proper notations for sound exposure level using this reference are $L_{W/(400\,\mu\text{Pa}^2\cdot\text{s})}$ or L_W (re 400 μPa²·s), but the notations dB SEL, dB(SEL), dBSEL, or dB_{SEL} are very common, even if they are not accepted by the SI.

Sound Intensity

Sound intensity also known as acoustic intensity is defined as the sound power per unit area. The SI unit of intensity, which includes sound intensity, is the watt per square meter (W/m²). One application is the noise measurement of sound intensity in the air at a listener's location as a sound energy quantity.

Sound intensity is not the same physical quantity as sound pressure. Hearing is directly sensitive to sound pressure which is related to sound intensity. In consumer audio electronics, the level differences are called "intensity" differences, but sound intensity is a specifically defined quantity and cannot be sensed by a simple microphone. The rate at which sound energy passes through a unit area held perpendicular to the direction of propagation of sound waves is called intensity of sound.

Mathematical Definition

Sound intensity, denoted I, is defined by

$$\mathbf{I} = p\mathbf{v}$$

where

- p is the sound pressure;
- v is the particle velocity.

Both I and v are vectors, which means that both have a *direction* as well as a magnitude. The direction of sound intensity is the average direction in which energy is flowing.

The average sound intensity during time T is given by

$$\langle \mathbf{I} \rangle = \frac{1}{T}\int_0^T p(t)\mathbf{v}(t)\mathrm{d}t.$$

Inverse-square Law

For a *spherical* sound wave, the intensity in the radial direction as a function of distance r from the centre of the sphere is given by

$$I(r) = \frac{P}{A(r)} = \frac{P}{4\pi r^2},$$

where

- P is the sound power;
- $A(r)$ is the area of a sphere of radius r.

Thus sound intensity decreases as $1/r^2$ from the centre of the sphere:

$$I(r) \propto \frac{1}{r^2}.$$

This relationship is an *inverse-square law.*

Sound Intensity Level

Sound intensity level (SIL) or acoustic intensity level is the level (a logarithmic quantity) of the intensity of a sound relative to a reference value. It is denoted L_I, expressed in dB, and defined by

$$L_I = \frac{1}{2}\ln\left(\frac{I}{I_0}\right)\mathrm{Np} = \log_{10}\left(\frac{I}{I_0}\right)\mathrm{B} = 10\log_{10}\left(\frac{I}{I_0}\right)\mathrm{dB},$$

where

- I is the sound intensity;
- I_0 is the *reference sound intensity*;
- 1 Np = 1 is the neper;
- 1 B = (1/2) ln(10) is the bel;
- 1 dB = (1/20) ln(10) is the decibel.

The commonly used reference sound intensity in air is

$$I_0 = 1\,\text{pW/m}^2.$$

The proper notations for sound intensity level using this reference are $L_{I/(1\,\text{pW/m}^2)}$ or L_I (re 1 pW/m²), but the notations dB SIL, dB(SIL), dBSIL, or dB_{SIL} are very common, even if they are not accepted by the SI.

The reference sound intensity I_0 is defined such that a progressive plane wave has the same value of sound intensity level (SIL) and sound pressure level (SPL), since

$$I \propto p^2.$$

The equality of SIL and SPL requires that

$$\frac{I}{I_0} = \frac{p^2}{p_0^2},$$

where p_0 = 20 μPa is the reference sound pressure.

For a *progressive* spherical wave,

$$\frac{p}{v} = z_0,$$

where z_0 is the characteristic specific acoustic impedance. Thus,

$$I_0 = \frac{p_0^2 I}{p^2} = \frac{p_0^2 pv}{p^2} = \frac{p_0^2}{z_0}.$$

In air at ambient temperature, z_0 = 410 Pa·s/m, hence the reference value I_0 = 1 pW/m².

In an anechoic chamber, which approximates a free field (no reflection), the SIL can be taken as being equal to the SPL. This fact is exploited to measure sound power in anechoic conditions.

Measurement

One method of sound intensity measurement involves the use of two microphones located close to each other, normal to the direction of sound energy flow. A signal analyser is used to compute the crosspower between the measured pressures and the sound intensity is derived from (proportional to) the imaginary part of the crosspower.

Speed of Sound

The speed of sound is the distance travelled per unit time by a sound wave as it propagates through an elastic medium. In dry air at 20 °C (68 °F), the speed of sound is 343.2 metres per second (1,126 ft/s; 1,236 km/h; 768 mph; 667 kn), or a kilometre in 2.914 s or a mile in 4.689 s.

The speed of sound in an ideal gas depends only on its temperature and composition. The speed has a weak dependence on frequency and pressure in ordinary air, deviating slightly from ideal behavior.

In common everyday speech, *speed of sound* refers to the speed of sound waves in air. However, the speed of sound varies from substance to substance: sound travels most slowly in gases; it travels faster in liquids; and faster still in solids. For example (as noted above), sound travels at 343.2 m/s in air; it travels at 1,484 m/s in water (4.3 times as fast as in air); and at 5,120 m/s in iron. In an exceptionally stiff material such as diamond, sound travels at 12,000 m/s; which is around the maximum speed that sound will travel under normal conditions.

Sound waves in solids are composed of compression waves (just as in gases and liquids), but there is also a different type of sound wave called a shear wave, which occurs only in solids. These different types of waves in solids usually travel at different speeds, as exhibited in seismology. The speed of a compression sound wave in solids is determined by the medium's compressibility, shear modulus and density. The speed of shear waves is determined only by the solid material's shear modulus and density.

In fluid dynamics, the speed of sound in a fluid medium (gas or liquid) is used as a relative measure for the speed of an object moving through the medium. The ratio of the speed of an object to the speed of sound in the fluid is called the object's Mach number. Objects moving at speeds greater than *Mach1* are said to be traveling at supersonic speeds.

History

Sir Isaac Newton computed the speed of sound in air as 979 feet per second (298 m/s), which is too low by about 15%, but had neglected the effect of fluctuating temperature; that was later rectified by Laplace.

During the 17th century, there were several attempts to measure the speed of sound accurately, including attempts by Marin Mersenne in 1630 (1,380 Parisian feet per second), Pierre Gassendi in 1635 (1,473 Parisian feet per second) and Robert Boyle (1,125 Parisian feet per second).

In 1709, the Reverend William Derham, Rector of Upminster, published a more accurate measure of the speed of sound, at 1,072 Parisian feet per second. Derham used a telescope from the tower of the church of St Laurence, Upminster to observe the flash of a distant shotgun being fired, and then measured the time until he heard the gunshot with a half second pendulum. Measurements were made of gunshots from a number of local landmarks, including North Ockendon church. The distance was known by triangulation, and thus the speed that the sound had travelled was calculated.

Basic Concept

The transmission of sound can be illustrated by using a model consisting of an array of balls interconnected by springs. For real material the balls represent molecules and the springs represent the bonds between them. Sound passes through the model by compressing and expanding the springs, transmitting energy to neighbouring balls, which transmit energy to *their* springs, and so on. The speed of sound through the model depends on the stiffness of the springs, and the mass of the balls. As long as the spacing of the balls remains constant, stiffer springs transmit energy more quickly, and more massive balls transmit energy more slowly. Effects like dispersion and reflection can also be understood using this model.

In a real material, the stiffness of the springs is called the elastic modulus, and the mass corresponds to the density. All other things being equal (ceteris paribus), sound will travel more slowly in spongy materials, and faster in stiffer ones. For instance, sound will travel 1.59 times faster in nickel than in bronze, due to the greater stiffness of nickel at about the same density. Similarly, sound travels about 1.41 times faster in light hydrogen (protium) gas than in heavy hydrogen (deuterium) gas, since deuterium has similar properties but twice the density. At the same time, "compression-type" sound will travel faster in solids than in liquids, and faster in liquids than in gases, because the solids are more difficult to compress than liquids, while liquids in turn are more difficult to compress than gases.

Some textbooks mistakenly state that the speed of sound increases with increasing density. This is usually illustrated by presenting data for three materials, such as air, water and steel, which also have vastly different compressibilities which more than make up for the density differences. An illustrative example of the two effects is that sound travels only 4.3 times faster in water than air, despite enormous differences in compressibility of the two media. The reason is that the larger density of water, which works to *slow* sound in water relative to air, nearly makes up for the compressibility differences in the two media.

Compression and Shear Waves

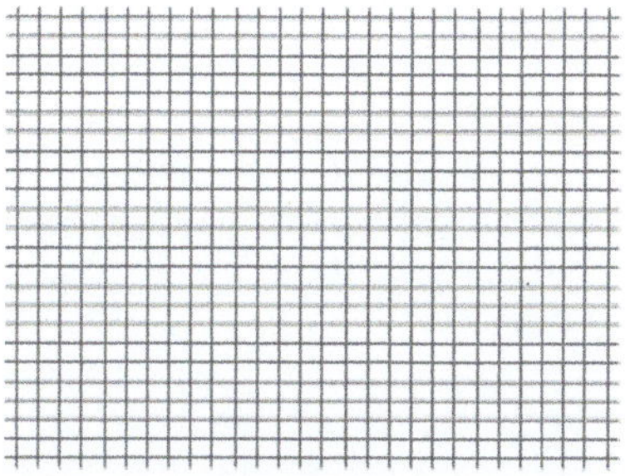

Pressure-pulse or compression-type wave (longitudinal wave) confined to a plane. This is the only type of sound wave that travels in fluids (gases and liquids). A pressure-type wave may also travel in solids, along with other types of waves (transverse waves)

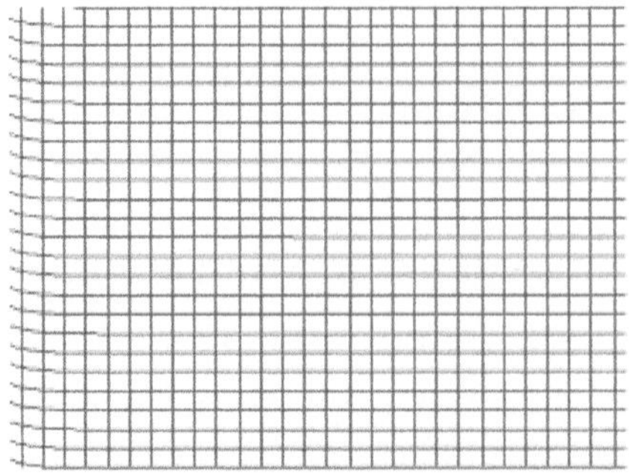

Transverse wave affecting atoms initially confined to a plane. This additional type of sound wave (additional type of elastic wave) travels only in solids, for it requires a sideways shearing motion which is supported by the presence of elasticity in the solid. The sideways shearing motion may take place in **any** direction which is at right-angle to the direction of wave-travel (only one shear direction is shown here, at right angles to the plane). Furthermore, the right-angle shear direction may change over time and distance, resulting in different types of *polarization* of shear-waves

In a gas or liquid, sound consists of compression waves. In solids, waves propagate as two different types. A longitudinal wave is associated with compression and decompression in the direction of travel, and is the same process in gases and liquids, with an analogous compression-type wave in solids. Only compression waves are supported in gases and liquids. An additional type of wave, the transverse wave, also called a shear wave, occurs only in solids because only solids support elastic deformations. It is due to elastic deformation of the medium perpendicular to the direction of wave travel; the direction of shear-deformation is called the "polarization" of this type of wave. In general, transverse waves occur as a pair of orthogonal polarizations.

These different waves (compression waves and the different polarizations of shear waves) may have different speeds at the same frequency. Therefore, they arrive at an observer at different times, an extreme example being an earthquake, where sharp compression waves arrive first, and rocking transverse waves seconds later.

The speed of a compression wave in fluid is determined by the medium's compressibility and density. In solids, the compression waves are analogous to those in fluids, depending on compressibility, density, and the additional factor of shear modulus. The speed of shear waves, which can occur only in solids, is determined simply by the solid material's shear modulus and density.

Equations

The speed of sound in mathematical notation is conventionally represented by c, from the Latin *celeritas* meaning "velocity".

In general, the speed of sound *c* is given by the Newton–Laplace equation:

$$c = \sqrt{\frac{K_s}{\rho}},$$

where

- K_s is a coefficient of stiffness, the isentropic bulk modulus (or the modulus of bulk elasticity for gases);
- ρ is the density.

Thus the speed of sound increases with the stiffness (the resistance of an elastic body to deformation by an applied force) of the material, and decreases with the density. For ideal gases the bulk modulus K is simply the gas pressure multiplied by the dimensionless adiabatic index, which is about 1.4 for air under normal conditions of pressure and temperature.

For general equations of state, if classical mechanics is used, the speed of sound c is given by

$$c = \sqrt{\left(\frac{\partial p}{\partial \rho}\right)_s},$$

where

- p is the pressure;
- ρ is the density and the derivative is taken isentropically, that is, at constant entropy s.

If relativistic effects are important, the speed of sound is calculated from the relativistic Euler equations.

In a non-dispersive medium, the speed of sound is independent of sound frequency, so the speeds of energy transport and sound propagation are the same for all frequencies. Air, a mixture of oxygen and nitrogen, constitutes a non-dispersive medium. However, air does contain a small amount of CO_2 which *is* a dispersive medium, and causes dispersion to air at ultrasonic frequencies (> 28 kHz).

In a dispersive medium, the speed of sound is a function of sound frequency, through the dispersion relation. Each frequency component propagates at its own speed, called the phase velocity, while the energy of the disturbance propagates at the group velocity. The same phenomenon occurs with light waves.

Dependence on the Properties of the Medium

The speed of sound is variable and depends on the properties of the substance through which the wave is travelling. In solids, the speed of transverse (or shear) waves depends on the shear deformation under shear stress (called the shear modulus), and the density of the medium. Longitudinal (or compression) waves in solids depend on the same two factors with the addition of a dependence on compressibility.

In fluids, only the medium's compressibility and density are the important factors, since fluids do not transmit shear stresses. In heterogeneous fluids, such as a liquid filled with gas bubbles, the density of the liquid and the compressibility of the gas affect the speed of sound in an additive manner, as demonstrated in the hot chocolate effect.

In gases, adiabatic compressibility is directly related to pressure through the heat capacity ratio (adiabatic index), while pressure and density are inversely related to the temperature and molecular weight, thus making only the completely independent properties of *temperature and molecular structure* important (heat capacity ratio may be determined by temperature and molecular structure, but simple molecular weight is not sufficient to determine it).

In low molecular weight gases such as helium, sound propagates faster as compared to heavier gases such as xenon. For monatomic gases, the speed of sound is about 75% of the mean speed that the atoms move in that gas.

For a given ideal gas the molecular composition is fixed, and thus the speed of sound depends only on its temperature. At a constant temperature, the gas pressure has no effect on the speed of sound, since the density will increase, and since pressure and density (also proportional to pressure) have equal but opposite effects on the speed of sound, and the two contributions cancel out exactly. In a similar way, compression waves in solids depend both on compressibility and density—just as in liquids—but in gases the density contributes to the compressibility in such a way that some part of each attribute factors out, leaving only a dependence on temperature, molecular weight, and heat capacity ratio which can be independently derived from temperature and molecular composition. Thus, for a single given gas (assuming the molecular weight does not change) and over a small temperature range (for which the heat capacity is relatively constant), the speed of sound becomes dependent on only the temperature of the gas.

In non-ideal gas behavior regimen, for which the van der Waals gas equation would be used, the proportionality is not exact, and there is a slight dependence of sound velocity on the gas pressure.

Humidity has a small but measurable effect on the speed of sound (causing it to increase by about 0.1%–0.6%), because oxygen and nitrogen molecules of the air are replaced by lighter molecules of water. This is a simple mixing effect.

Altitude Variation and Implications for Atmospheric Acoustics

In the Earth's atmosphere, the chief factor affecting the speed of sound is the temperature. For a given ideal gas with constant heat capacity and composition, the speed of sound is dependent *solely* upon temperature. In such an ideal case, the effects of decreased density and decreased pressure of altitude cancel each other out, save for the residual effect of temperature.

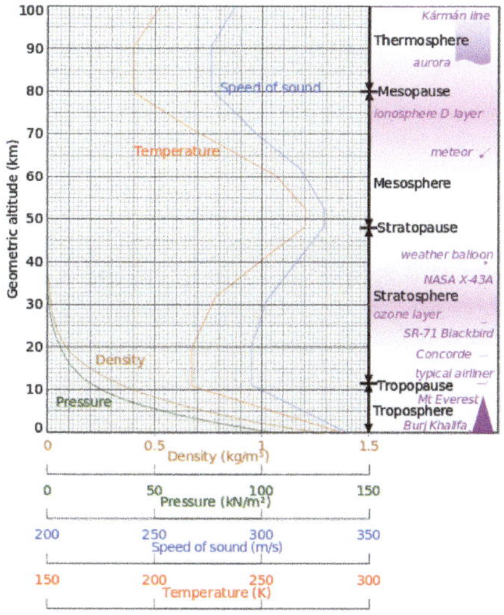

Density and pressure decrease smoothly with altitude, but temperature (red) does not. The speed of sound (blue) depends only on the complicated temperature variation at altitude and can be calculated from it, since isolated density and pressure effects on the speed of sound cancel each other. Speed of sound increases with height in two regions of the stratosphere and thermosphere, due to heating effects in these regions.

Since temperature (and thus the speed of sound) decreases with increasing altitude up to 11 km, sound is refracted upward, away from listeners on the ground, creating an acoustic shadow at some distance from the source. The decrease of the speed of sound with height is referred to as a negative sound speed gradient.

However, there are variations in this trend above 11 km. In particular, in the stratosphere above about 20 km, the speed of sound increases with height, due to an increase in temperature from heating within the ozone layer. This produces a positive speed of sound gradient in this region. Still another region of positive gradient occurs at very high altitudes, in the aptly-named thermosphere above 90 km.

Practical Formula for Dry Air

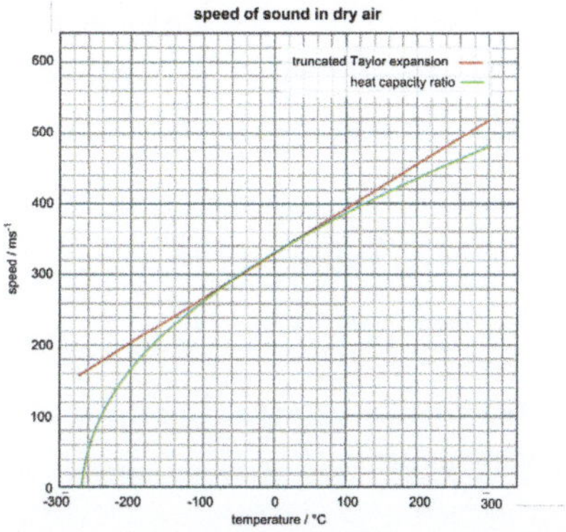

Approximation of the speed of sound in dry air based on the heat capacity ratio (in green) against the truncated Taylor expansion (in red).

The approximate speed of sound in dry (0% humidity) air, in meters per second, at temperatures near 0 °C, can be calculated from

$$c_{\text{air}} = (331.3 + 0.606 \cdot \vartheta) \text{ m/s},$$

where ϑ is the temperature in degrees Celsius (°C).

This equation is derived from the first two terms of the Taylor expansion of the following more accurate equation:

$$c_{\text{air}} = 331.3\sqrt{1 + \frac{\vartheta}{273.15}} \text{ m/s}.$$

Dividing the first part, and multiplying the second part, on the right hand side, by √273.15 gives the exactly equivalent form

$$c_{\text{air}} = 20.05\sqrt{\vartheta + 273.15} \text{ m/s}.$$

The value of 331.3 m/s, which represents the speed at 0 °C (or 273.15 K), is based on theoretical (and some measured) values of the heat capacity ratio, γ, as well as on the fact that at 1 atm real air is very well described by the ideal gas approximation. Commonly found values for the speed of sound at 0 °C may vary from 331.2 to 331.6 due to the assumptions made when it is calculated. If ideal gas γ is assumed to be 7/5 = 1.4 exactly, the 0 °C speed is calculated to be 331.3 m/s, the coefficient used above.

This equation is correct to a much wider temperature range, but still depends on the approximation of heat capacity ratio being independent of temperature, and for this reason will fail, particularly at higher temperatures. It gives good predictions in relatively dry, cold, low pressure conditions, such as the Earth's stratosphere. The equation fails at extremely low pressures and short wavelengths, due to dependence on the assumption that the wavelength of the sound in the gas is much longer than the average mean free path between gas molecule collisions. A derivation of these equations will be given in the following section.

A graph comparing results of the two equations is at right, using the slightly different value of 331.5 m/s for the speed of sound at 0 °C.

Details

Speed of Sound in Ideal Gases and Air

For an ideal gas, K (the bulk modulus in equations above, equivalent to C, the coefficient of stiffness in solids) is given by

$$K = \gamma \cdot p,$$

thus, from the Newton–Laplace equation above, the speed of sound in an ideal gas is given by

$$c = \sqrt{\gamma \cdot \frac{p}{\rho}},$$

where

- γ is the adiabatic index also known as the *isentropic expansion factor*. It is the ratio of specific heats of a gas at a constant-pressure to a gas at a constant-volume(C_p / C_v), and arises because a classical sound wave induces an adiabatic compression, in which the heat of the compression does not have enough time to escape the pressure pulse, and thus contributes to the pressure induced by the compression;
- p is the pressure;
- ρ is the density.

Using the ideal gas law to replace p with nRT/V, and replacing ρ with nM/V, the equation for an ideal gas becomes

$$c_{\text{ideal}} = \sqrt{\gamma \cdot \frac{p}{\rho}} = \sqrt{\frac{\gamma \cdot R \cdot T}{M}} = \sqrt{\frac{\gamma \cdot k \cdot T}{m}},$$

where

- c_{ideal} is the speed of sound in an ideal gas;
- R (approximately 8.314,5 $\text{J} \cdot \text{mol}^{-1} \cdot \text{K}^{-1}$) is the molar gas constant(universal gas constant);
- k is the Boltzmann constant;
- γ (gamma) is the adiabatic index. At room temperature, where thermal energy is fully partitioned into rotation (rotations are fully excited) but quantum effects prevent excitation of vibrational modes, the value is 7/5 = 1.400 for diatomic molecules, according to kinetic theory. Gamma is actually experimentally measured over a range from 1.399,1 to 1.403 at 0 °C, for air. Gamma is exactly 5/3 = 1.6667 for monatomic gases such as noble gases;
- T is the absolute temperature;
- M is the molar mass of the gas. The mean molar mass for dry air is about 0.028,964,5 kg/mol;
- n is the number of moles;
- m is the mass of a single molecule.

This equation applies only when the sound wave is a small perturbation on the ambient condition, and the certain other noted conditions are fulfilled, as noted below. Calculated values for c_{air} have been found to vary slightly from experimentally determined values.

Newton famously considered the speed of sound before most of the development of thermodynamics and so incorrectly used isothermal calculations instead of adiabatic. His result was missing the factor of γ but was otherwise correct.

Numerical substitution of the above values gives the ideal gas approximation of sound velocity for gases, which is accurate at relatively low gas pressures and densities (for air, this includes standard Earth sea-level conditions). Also, for diatomic gases the use of $\gamma = 1.400,0$ requires that the gas exists in a temperature range high enough that rotational heat capacity is fully excited (i.e., molecular rotation is fully used as a heat energy "partition" or reservoir); but at the same time the temperature must be low enough that molecular vibrational modes contribute no heat capacity (i.e., insignificant heat goes into vibration, as all vibrational quantum modes above the minimum-energy-mode, have energies too high to be populated by a significant number of molecules at this temperature). For air, these conditions are fulfilled at room temperature, and also temperatures considerably below room temperature.

For air, we use a simplified symbol

$$R_* = R / M_{air}.$$

Additionally, if temperatures in degrees Celsius(°C) are to be used to calculate air speed in the region near 273 kelvin, then Celsius temperature $\theta = T - 273.15$ may be used. Then

$$c_{ideal} = \sqrt{\gamma \cdot R_* \cdot T} = \sqrt{\gamma \cdot R_* \cdot (\vartheta + 273.15)},$$

$$c_{ideal} = \sqrt{\gamma \cdot R_* \cdot 273.15} \cdot \sqrt{1 + \frac{\vartheta}{273.15}}.$$

For dry air, where θ (theta) is the temperature in degrees Celsius(°C).

Making the following numerical substitutions,

$$R = 8.314,510\,\mathrm{J/(mol \cdot K)}$$

is the molar gas constant in J/mole/Kelvin, and

$$M_{air} = 0.028,964,5\,\mathrm{kg/mol}$$

is the mean molar mass of air, in kg; and using the ideal diatomic gas value of γ = 1.400,0.

Then

$$c_{\text{air}} = 331.3 \sqrt{1 + \frac{\vartheta^{\circ}\text{C}}{273.15^{\circ}\text{C}}} \ \text{m/s}.$$

Using the first two terms of the Taylor expansion:

$$c_{\text{air}} = 331.3 \ (1 + \frac{\vartheta^{\circ}\text{C}}{2 \cdot 273.15^{\circ}\text{C}}) \ \text{m/s},$$

$$c_{\text{air}} = (331.3 + 0.606^{\circ}\text{C}^{-1} \cdot \vartheta) \ \text{m/s}.$$

The derivation includes the first two equations given in the *Practical formula for dry air* section above.

Effects Due to Wind Shear

The speed of sound varies with temperature. Since temperature and sound velocity normally decrease with increasing altitude, sound is refracted upward, away from listeners on the ground, creating an acoustic shadow at some distance from the source. Wind shear of 4 m/(s · km) can produce refraction equal to a typical temperature lapse rate of 7.5 °C/km. Higher values of wind gradient will refract sound downward toward the surface in the downwind direction, eliminating the acoustic shadow on the downwind side. This will increase the audibility of sounds downwind. This downwind refraction effect occurs because there is a wind gradient; the sound is not being carried along by the wind.

For sound propagation, the exponential variation of wind speed with height can be defined as follows:

$$U(h) = U(0)h^{\zeta},$$

$$\frac{\mathrm{d}U}{\mathrm{d}H}(h) = \zeta \frac{U(h)}{h},$$

where

- $U(h)$ is the speed of the wind at height h;
- ζ is the exponential coefficient based on ground surface roughness, typically between 0.08 and 0.52;
- $\mathrm{d}U/\mathrm{d}H(h)$ is the expected wind gradient at height h.

In the 1862 American Civil War Battle of Iuka, an acoustic shadow, believed to have been enhanced by a northeast wind, kept two divisions of Union soldiers out of the battle, because they could not hear the sounds of battle only 10 km (six miles) downwind.

Tables

In the standard atmosphere:

- T_0 is 273.15 K (= 0 °C = 32 °F), giving a theoretical value of 331.3 m/s (= 1086.9 ft/s = 1193 km/h = 741.1 mph = 644.0 kn). Values ranging from 331.3-331.6 may be found in reference literature, however;
- T_{20} is 293.15 K (= 20 °C = 68 °F), giving a value of 343.2 m/s (= 1126.0 ft/s = 1236 km/h = 767.8 mph = 667.2 kn);
- T_{25} is 298.15 K (= 25 °C = 77 °F), giving a value of 346.1 m/s (= 1135.6 ft/s = 1246 km/h = 774.3 mph = 672.8 kn).

In fact, assuming an ideal gas, the speed of sound *c* depends on temperature only, not on the pressure or density (since these change in lockstep for a given temperature and cancel out). Air is almost an ideal gas. The temperature of the air varies with altitude, giving the following variations in the speed of sound using the standard atmosphere—*actual conditions may vary*.

Effect of temperature on properties of air			
Temperature T (°C)	**Speed of sound c (m/s)**	**Density of air ρ (kg/m³)**	**Characteristic specific acoustic impedance z_0 (Pa·s/m)**
35	351.88	1.1455	403.2
30	349.02	1.1644	406.5
25	346.13	1.1839	409.4
20	343.21	1.2041	413.3
15	340.27	1.2250	416.9
10	337.31	1.2466	420.5
5	334.32	1.2690	424.3
0	331.30	1.2922	428.0
−5	328.25	1.3163	432.1
−10	325.18	1.3413	436.1
−15	322.07	1.3673	440.3
−20	318.94	1.3943	444.6
−25	315.77	1.4224	449.1

Given normal atmospheric conditions, the temperature, and thus speed of sound, varies with altitude:

Altitude	Temperature	m/s	km/h	mph	kn
Sea level	15 °C (59 °F)	340	1,225	761	661

11,000 m–20,000 m (Cruising altitude of commercial jets, and first supersonic flight)	−57 °C (−70 °F)	295	1,062	660	573
29,000 m (Flight of X-43A)	−48 °C (−53 °F)	301	1,083	673	585

Effect of Frequency and Gas Composition

General Physical Considerations

The medium in which a sound wave is travelling does not always respond adiabatically, and as a result the speed of sound can vary with frequency.

The limitations of the concept of speed of sound due to extreme attenuation are also of concern. The attenuation which exists at sea level for high frequencies applies to successively lower frequencies as atmospheric pressure decreases, or as the mean free path increases. For this reason, the concept of speed of sound (except for frequencies approaching zero) progressively loses its range of applicability at high altitudes. The standard equations for the speed of sound apply with reasonable accuracy only to situations in which the wavelength of the soundwave is considerably longer than the mean free path of molecules in a gas.

The molecular composition of the gas contributes both as the mass (M) of the molecules, and their heat capacities, and so both have an influence on speed of sound. In general, at the same molecular mass, monatomic gases have slightly higher speed of sound (over 9% higher) because they have a higher γ (5/3 = 1.66...) than diatomics do (7/5 = 1.4). Thus, at the same molecular mass, the speed of sound of a monatomic gas goes up by a factor of

$$\frac{c_{\text{gas,monatomic}}}{c_{\text{gas,diatomic}}} = \sqrt{\frac{5/3}{7/5}} = \sqrt{\frac{25}{21}} = 1.091\ldots$$

This gives the 9% difference, and would be a typical ratio for speeds of sound at room temperature in helium vs. deuterium, each with a molecular weight of 4. Sound travels faster in helium than deuterium because adiabatic compression heats helium more, since the helium molecules can store heat energy from compression only in translation, but not rotation. Thus helium molecules (monatomic molecules) travel faster in a sound wave and transmit sound faster. (Sound travels at about 70% of the mean molecular speed in gases; the figure is 75% in monatomic gases and 68% in diatomic gases).

Note that in this example we have assumed that temperature is low enough that heat capacities are not influenced by molecular vibration. However, vibrational modes simply cause gammas which decrease toward 1, since vibration modes in a polyatomic gas gives the gas additional ways to store heat which do not affect temperature, and thus do not affect molecular velocity and sound velocity. Thus, the effect of higher temperatures and vibrational heat capacity acts to increase the difference between the speed of sound in monatomic vs. polyatomic molecules, with the speed remaining greater in monatomics.

Practical Application to Air

By far the most important factor influencing the speed of sound in air is temperature. The speed is proportional to the square root of the absolute temperature, giving an increase of about 0.6 m/s per degree Celsius. For this reason, the pitch of a musical wind instrument increases as its temperature increases.

The speed of sound is raised by humidity but decreased by carbon dioxide. The difference between 0% and 100% humidity is about 1.5 m/s at standard pressure and temperature, but the size of the humidity effect increases dramatically with temperature. The carbon dioxide content of air is not fixed, due to both carbon pollution and human breath (e.g., in the air blown through wind instruments).

The dependence on frequency and pressure are normally insignificant in practical applications. In dry air, the speed of sound increases by about 0.1 m/s as the frequency rises from 10 Hz to 100 Hz. For audible frequencies above 100 Hz it is relatively constant. Standard values of the speed of sound are quoted in the limit of low frequencies, where the wavelength is large compared to the mean free path.

Mach Number

U.S. Navy F/A-18 traveling near the speed of sound. The white halo consists of condensed water droplets formed by the sudden drop in air pressure behind the shock cone around the aircraft.

Mach number, a useful quantity in aerodynamics, is the ratio of air speed to the local speed of sound. At altitude, for reasons explained, Mach number is a function of temperature. Aircraft flight instruments, however, operate using pressure differential to compute Mach number, not temperature. The assumption is that a particular pressure represents a particular altitude and, therefore, a standard temperature. Aircraft flight instruments need to operate this way because the stagnation pressure sensed by a Pitot tube is dependent on altitude as well as speed.

Experimental Methods

A range of different methods exist for the measurement of sound in air.

The earliest reasonably accurate estimate of the speed of sound in air was made by William Derham, and acknowledged by Isaac Newton. Derham had a telescope at the top of the tower of the Church of St Laurence in Upminster, England. On a calm day, a synchronized pocket watch would be given to an assistant who would fire a shotgun at a pre-determined time from a conspicuous point some miles away, across the countryside. This could be confirmed by telescope. He then measured the interval between seeing gunsmoke and arrival of the sound using a half-second pendulum. The distance from where the gun was fired was found by triangulation, and simple division (distance/time) provided velocity. Lastly, by making many observations, using a range of different distances, the inaccuracy of the half-second pendulum could be averaged out, giving his final estimate of the speed of sound. Modern stopwatches enable this method to be used today over distances as short as 200–400 meters, and not needing something as loud as a shotgun.

Single-shot Timing Methods

The simplest concept is the measurement made using two microphones and a fast recording device such as a digital storage scope. This method uses the following idea.

If a sound source and two microphones are arranged in a straight line, with the sound source at one end, then the following can be measured:

1. The distance between the microphones (x), called microphone basis.
2. The time of arrival between the signals (delay) reaching the different microphones (t).

Then $v = x/t$.

Other Methods

In these methods the time measurement has been replaced by a measurement of the inverse of time (frequency).

Kundt's tube is an example of an experiment which can be used to measure the speed of sound in a small volume. It has the advantage of being able to measure the speed of sound in any gas. This method uses a powder to make the nodes and antinodes visible to the human eye. This is an example of a compact experimental setup.

A tuning fork can be held near the mouth of a long pipe which is dipping into a barrel of water. In this system it is the case that the pipe can be brought to resonance if the length of the air column in the pipe is equal to $(1 + 2n)\lambda/4$ where n is an integer. As the antinodal point for the pipe at the open end is slightly outside the mouth of the pipe it is best to find two or more points of resonance and then measure half a wavelength between these.

Here it is the case that $v = f\lambda$.

High-precision Measurements in Air

The effect from impurities can be significant when making high-precision measurements. Chemical desiccants can be used to dry the air, but will in turn contaminate the sample. The air can be dried cryogenically, but this has the effect of removing the carbon dioxide as well; therefore many high-precision measurements are performed with air free of carbon dioxide rather than with natural air. A 2002 review found that a 1963 measurement by Smith and Harlow using a cylindrical resonator gave "the most probable value of the standard speed of sound to date." The experiment was done with air from which the carbon dioxide had been removed, but the result was then corrected for this effect so as to be applicable to real air. The experiments were done at 30 °C but corrected for temperature in order to report them at 0 °C. The result was 331.45 ± 0.01 m/s for dry air at STP, for frequencies from 93 Hz to 1,500 Hz.

Non-gaseous Media

Speed of Sound in Solids

Three-dimensional Solids

In a solid, there is a non-zero stiffness both for volumetric deformations and shear deformations. Hence, it is possible to generate sound waves with different velocities dependent on the deformation mode. Sound waves generating volumetric deformations (compression) and shear deformations (shearing) are called pressure waves (longitudinal waves) and shear waves (transverse waves), respectively. In earthquakes, the corresponding seismic waves are called P-waves (primary waves) and S-waves (secondary waves), respectively. The sound velocities of these two types of waves propagating in a homogeneous 3-dimensional solid are respectively given by

$$_{\text{solid,p}} = \sqrt{\frac{K \quad - G}{\rho}} = \sqrt{\frac{(1 \quad)}{\rho(1+\nu)(1-2\nu)}}$$

$$c_{\text{solid,s}} = \sqrt{\frac{G}{\rho}},$$

where

- K is the bulk modulus of the elastic materials;
- G is the shear modulus of the elastic materials;
- E is the Young's modulus;

- ρ is the density;
- v is Poisson's ratio.

The last quantity is not an independent one, as E = 3K(1 − 2v). Note that the speed of pressure waves depends both on the pressure and shear resistance properties of the material, while the speed of shear waves depends on the shear properties only.

Typically, pressure waves travel faster in materials than do shear waves, and in earthquakes this is the reason that the onset of an earthquake is often preceded by a quick upward-downward shock, before arrival of waves that produce a side-to-side motion. For example, for a typical steel alloy, K = 170 GPa, G = 80 GPa and ρ = 7,700 kg/m^3, yielding a compressional speed $c_{solid,p}$ of 6,000 m/s. This is in reasonable agreement with $c_{solid,p}$ measured experimentally at 5,930 m/s for a (possibly different) type of steel. The shear speed $c_{solid,s}$ is estimated at 3,200 m/s using the same numbers.

One-dimensional Solids

The speed of sound for pressure waves in stiff materials such as metals is sometimes given for "long rods" of the material in question, in which the speed is easier to measure. In rods where their diameter is shorter than a wavelength, the speed of pure pressure waves may be simplified and is given by:

$$c_{solid} = \sqrt{\frac{E}{\rho}},$$

where E is the Young's modulus. This is similar to the expression for shear waves, save that Young's modulus replaces the shear modulus. This speed of sound for pressure waves in long rods will always be slightly less than the same speed in homogeneous 3-dimensional solids, and the ratio of the speeds in the two different types of objects depends on Poisson's ratio for the material.

Speed of Sound in Liquids

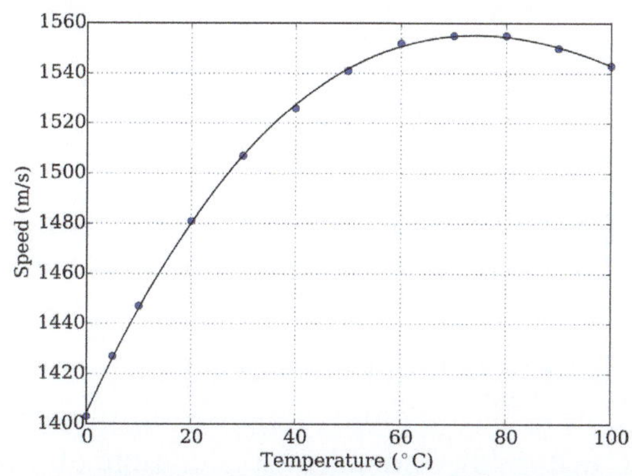

Speed of sound in water vs temperature.

In a fluid the only non-zero stiffness is to volumetric deformation (a fluid does not sustain shear forces).

Hence the speed of sound in a fluid is given by

$$c_{\text{fluid}} = \sqrt{\frac{K}{\rho}},$$

where K is the bulk modulus of the fluid.

Water

In fresh water, sound travels at about 1481 m/s at 20 °C. Speed of Sound in Pure Water for an online calculator. Applications of underwater sound can be found in sonar, acoustic communication and acoustical oceanography.

Seawater

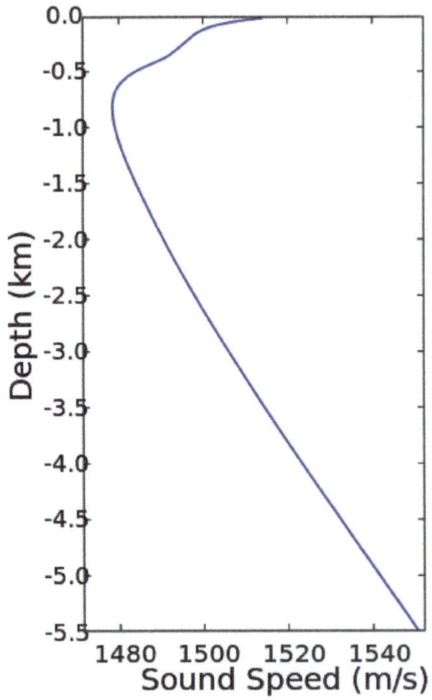

Speed of sound as a function of depth at a position north of Hawaii in the Pacific Ocean derived from the 2005 World Ocean Atlas. The SOFAR channel is centred on the minimum in the speed of sound at ca. 750-m depth.

In salt water that is free of air bubbles or suspended sediment, sound travels at about 1500 m/s. The speed of sound in seawater depends on pressure (hence depth), temperature (a change of 1 °C ~ 4 m/s), and salinity (a change of 1‰ ~ 1 m/s), and empirical equations have been derived to accurately calculate the speed of sound from these variables. Other factors affecting the speed of sound are minor. Since temperature decreases with depth while pressure and generally salinity increase, the profile of the speed of sound with depth generally shows a characteristic curve which decreases to a minimum at a depth of several hundred meters, then increases again with increasing depth (right).

A simple empirical equation for the speed of sound in sea water with reasonable accuracy for the world's oceans is due to Mackenzie:

$$c(T,S,z) = a_1 + a_2T + a_3T^2 + a_4T^3 + a_5(S-35) + a_6z + a_7z^2 + a_8T(S-35) + a_9Tz^3,$$

where

- T is the temperature in degrees Celsius;
- S is the salinity in parts per thousand;
- z is the depth in meters.

The constants $a_1, a_2, \ldots, a_9$ are

$$\begin{aligned} a_1 &= 1{,}448.96, & a_2 &= 4.591, & a_3 &= -5.304\times10^{-2}, \\ a_4 &= 2.374\times10^{-4}, & a_5 &= 1.340, & a_6 &= 1.630\times10^{-2}, \\ a_7 &= 1.675\times10^{-7}, & a_8 &= -1.025\times10^{-2}, & a_9 &= -7.139\times10^{-13}, \end{aligned}$$

with check value 1550.744 m/s for T = 25 °C, S = 35 parts per thousand, z = 1,000 m. This equation has a standard error of 0.070 m/s for salinity between 25 and 40 ppt. Speed of Sound in Sea-Water for an online calculator.

Other equations for the speed of sound in sea water are accurate over a wide range of conditions, but are far more complicated, e.g., that by V. A. Del Grosso and the Chen-Millero-Li Equation.

Speed of Sound in Plasma

The speed of sound in a plasma for the common case that the electrons are hotter than the ions (but not too much hotter) is given by the formula.

$$c_s = (\gamma ZkT_e / m_i)^{1/2} = 9.79\times10^3(\gamma ZT_e / \mu)^{1/2} \text{ m/s},$$

where

- m_i is the ion mass;
- μ is the ratio of ion mass to proton mass $\mu = m_i/m_p$;
- T_e is the electron temperature;

- Z is the charge state;
- k is Boltzmann constant;
- γ is the adiabatic index.

In contrast to a gas, the pressure and the density are provided by separate species, the pressure by the electrons and the density by the ions. The two are coupled through a fluctuating electric field.

Gradients

When sound spreads out evenly in all directions in three dimensions, the intensity drops in proportion to the inverse square of the distance. However, in the ocean there is a layer called the 'deep sound channel' or SOFAR channel which can confine sound waves at a particular depth.

In the SOFAR channel, the speed of sound is lower than that in the layers above and below. Just as light waves will refract towards a region of higher index, sound waves will refract towards a region where their speed is reduced. The result is that sound gets confined in the layer, much the way light can be confined in a sheet of glass or optical fiber. Thus, the sound is confined in essentially two dimensions. In two dimensions the intensity drops in proportion to only the inverse of the distance. This allows waves to travel much further before being undetectably faint.

A similar effect occurs in the atmosphere. Project Mogul successfully used this effect to detect a nuclear explosion at a considerable distance.

Noise Dosimeter

A noise dosimeter (American) or noise dosemeter (British) is a specialized sound level meter intended specifically to measure the noise exposure of a person integrated over a period of time; usually to comply with Health and Safety regulations such as the Occupational Safety and Health (OSHA) 29 CFR 1910.95 Occupational Noise Exposure Standard or EU Directive 2003/10/EC.

Noise Measurement

Noise dosimeters measure and store sound pressure levels (SPL) and, by integrating these measurements over time, provide a cumulative noise-exposure reading for a given period of time, such as an 8-hour workday. Dosimeters can function as personal or area noise monitors. In occupational settings, personal noise dosimeters are often worn on the body of a worker with the microphone mounted on the middle-top of the person's most exposed shoulder.

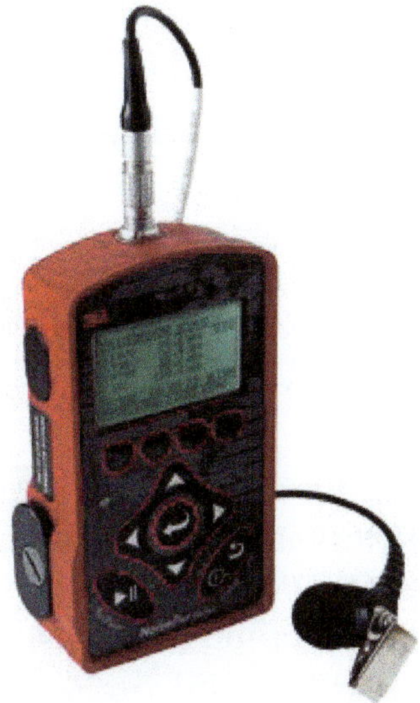

Fig 1 : 3M Noise Pro Dosimeter

Area monitoring can be used to estimate noise exposure when the noise levels are relatively constant and employees are not mobile. In workplaces where employees move about in different areas or where the noise intensity tends to fluctuate over time, noise exposure is generally more accurately estimated by the personal monitoring approach.

Dosimeters are also used to collect data for use in legal proceedings, development of engineering noise controls, and other industrial hygiene purposes. When planning to conduct noise exposure measurements, steps must be taken to ensure that the dosimeters are calibrated and operated according to manufacturers' specifications. It is also necessary to understand the properties of the acoustic environment, the main measurement objectives as they relate to determining the risk to hearing damage, and the limitations associated with the use of dosimeters. Dosimeter manufacturers recommend that the instrument be calibrated with an acoustical calibrator such as a pistonphone before and after each measurement to verify reliable operation. In addition to field calibration routines, the manufacturers recommend periodic comprehensive calibration and certification of the instrument by an accredited laboratory using traceable reference sources. Field calibration of contemporary dosimeters has been mostly automated through PC-based programs that run the calibration routine, document the time and date, and adjust for any offset in levels.

Occupational Settings

Current dosimeters are designed to provide the user with parameters such as noise dose, time-weighted average, sound exposure level, as well as peak, maximum, and minimum sound pressure levels. Most dosimeters also generate statistical and graphical representations of the collected data. ANSI S1.25 specifies that dosimeters should at least provide the following parameters:

Frequency weighting: A-weighting or C-weighting

Exponential averaging: F (fast); S (slow)

Criterion level: 90, 85, 84, 80, or V (variable)

Criterion duration: Hours

Threshold level: 90, 80, or V (variable)

Exchange rate: 5, 4, or 3

Standardization

The international body that specifies the technical requirements of such instruments as sound level meters and dosimeters is the International Electrotechnical Commission (IEC) based in Geneva; whereas the method of their use is normally given in an International Organization for Standardization (ISO) publication. In the U.S., the American National Standards Institute (ANSI) ANSI S1.25-1991 (R2007) specifies the performance characteristics of personal noise dosimeters.

Use of Dosimeters

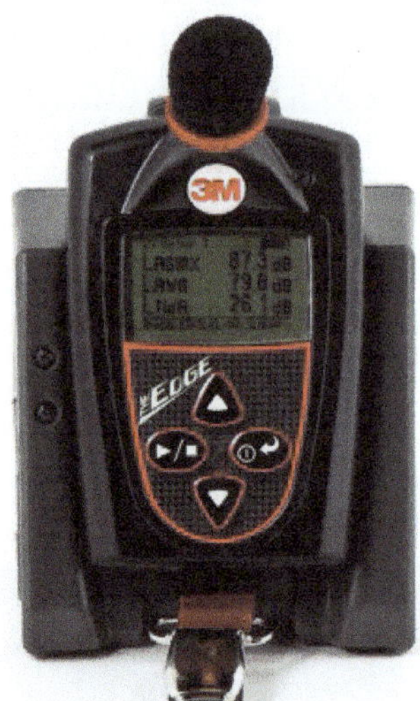

Fig 1 : 3M Edge 85 gram Dosimeter

The original dosimeters were designed to be belt worn with a microphone connected to the body of the dosimeter and mounted on the shoulder as near to the ear as practicable. These devices were worn for the full work shift and at the end would give a readout initially in percentage dose, or in some other exposure metric. These were the most common way of making measurements to meet legislation in the USA, but in Europe, the conventional sound level meter was favoured. There were many reasons for this, but in general in Europe the dosimeter was distrusted for several reasons, some being.

- The cable was considered dangerous as it could catch on rotating machinery

- The dosimeter could tell you the level had been exceeded, but it did not say when this happened
- Workers could falsify the data very easily
- The device was big enough to affect the work pattern

In the USA - where most of the early devices were manufactured, these reasons did not seem to matter so much.

To remove these European objections, dosimeters became smaller and started to include a data store where the Time History of the noise, usually in the form of Short Equivalent Sound Level (Leq) was stored. This data could be transferred to a personal computer and the exact pattern of the noise exposure minute by minute plotted. The usual method used was to store data in the form of Short Leq, a French concept that helped to bring computers into acoustics. As well, dosimeters started to incorporate besides the A-weighting a second C-frequency-weighted channel that allowed the true peak to be indicated. By the time the PSEM standard was published, many major sound level meter companies - in both Europe and the USA had a dosimeter in their range.

Companies

The following major manufacturers are among those who offer noise dosimeters:

- Denmark Brüel & Kjær
- United Kingdom Casella CEL Ltd.
- United Kingdom Cirrus Research plc
- United Kingdom Pulsar Instruments plc
- USA 3M
- USA Larson-Davis
- France 01dB-Metravib
- Poland SVANTEK

Ambient Noise Level

In atmospheric sounding and noise pollution, ambient noise level (sometimes called background noise level, reference sound level, or room noise level) is the background sound pressure level at a given location, normally specified as a reference level to study a new intrusive sound source.

Ambient sound levels are often measured in order to map sound conditions over a spatial regime to understand their variation with locale. In this case the product of the investigation is a sound level contour map. Alternatively ambient noise levels may be measured to provide a reference point for analyzing an intrusive sound to a given environment. For example, sometimes aircraft noise is studied by measuring ambient sound without presence of any overflights, and then studying the noise addition by measurement or computer simulation of overflight events. Or roadway noise is measured as ambient sound, prior to introducing a hypothetical noise barrier intended to reduce that ambient noise level.

Ambient noise level is measured with a sound level meter. It is usually measured in dB relative to a reference pressure of 0.00002 Pa, *i.e.,* 20 μPa (micropascals) in SI units. A pascal is a newton per square meter. The centimeter-gram-second system of units, the reference sound pressure for measuring ambient noise level is 0.0002 dyn/cm^2. Most frequently ambient noise levels are measured using a frequency weighting filter, the most common being the A-weighting scale, such that resulting measurements are denoted dB(A), or decibels on the A-weighting scale.

References

- Murdin, Paul (Dec 25, 2008). Full Meridian of Glory: Perilous Adventures in the Competition to Measure the Earth. Springer Science & Business Media. pp. 35–36. ISBN 9780387755342.
- Everest, F. (2001). The Master Handbook of Acoustics. New York: McGraw-Hill. pp. 262–263. ISBN 0-07-136097-2.
- Volland, Hans (1995). Handbook of Atmospheric Electrodynamics. Boca Raton: CRC Press. p. 22. ISBN 0-8493-8647-0.
- Cozens, Peter (2006). The Darkest Days of the War: the Battles of Iuka and Corinth. Chapel Hill: The University of North Carolina Press. ISBN 0-8078-5783-1.
- Bannon, Mike; Kaputa, Frank. "The Newton–Laplace Equation and Speed of Sound". Thermal Jackets. Retrieved 3 May 2015.
- "Noise Dosimeters, Handbook of Noise and Vibration Control". John Wiley & Sons, Inc. 2008-04-08. Retrieved 2014-12-04.
- "EU Directive 2003/10/EC (normally called the Physical Agents Directive)". European Union. 2013-02-15. Retrieved 2013-08-26.
- "OSHA 29 CFR 1910.95 Appendix G Occupational Noise Exposure Standard". Occupational Health and Safety Administration. 2011-03-03. Retrieved 2012-09-10.

4

Effects of Noise Pollution on Health

Noise pollution has certain effects on the environment as well as an individual. Some of these effects are hearing problems, cardiovascular issues, sleeping disorders and tinnitus. Animals also face problems because of noise pollution as the frequency of their hearing is higher. The effects may not be felt strongly on a day-to-day basis, but it can be very damaging in the long run. This section helps the readers in understanding the effects produced by noise pollution.

Noise-induced Hearing Loss

Noise-induced hearing loss (NIHL) is hearing impairment resulting from exposure to loud sound. People may have a loss of perception of a narrow range of frequencies, impaired cognitive perception of sound, or other impairment, including sensitivity to sound or ringing in the ears.

Hearing may deteriorate gradually from chronic and repeated noise exposure, such as loud music or background noise, or suddenly, from a short high intensity noise such as a gunshot or airhorn. In both types, loud sound overstimulates delicate hearing cells, leading to the permanent injury or death of the cells. Once lost, hearing cannot be restored in humans. When exposure to hazards such as noise occur at work and is associated with hearing loss, it is referred to as occupational hearing loss.

There are a variety of prevention strategies available to avoid or reduce hearing loss. If not prevented, hearing loss can be managed through assistive devices and cognitive therapies. Prevention is the best defense against noise-induced hearing loss. The best, first option for protecting hearing is lowering the volume of sound at its source. Secondly, limiting the time of exposure to loud noise can reduce injury. Finally, physical protection from the noise can reduce its impact. Government regulations are designed to limit occupational exposure to dangerously loud noise.

The largest burden of NIHL, has been through occupational exposures; however, noise-induced hearing loss can also be due to unsafe recreational, residential, social, and military service-related noise exposures. It is estimated that 15% of young people are exposed to sufficient leisure noises (i.e. concerts, sporting events, daily activities, personal listening devices, etc.) to cause NIHL. There is not a limited list of noise sources that can cause hearing loss. Rather, it is important to understand that exposure to excessively high decibel levels from any sound source over time, can cause hearing loss.

Signs and Symptoms

The first symptom of noise-induced hearing loss is usually difficulty hearing a conversation against a noisy background. The effect of hearing loss on speech perception has two components. The first component is the loss of audibility, which is something like a decrease in overall volume. Modern hearing aids compensate this loss with amplification. However, difficulty in understanding speech represents selective frequency loss for which hearing aids and amplification do not help. This is known by different names such as "distortion," "clarity loss," and "Signal-to-Noise-Ratio (SNR)-loss." Consonants, due to their higher frequency, seem to be lost first. For example, the letters "s" and "t" are the common letters that are difficult to hear for those with hearing loss due to them being our highest frequency sound in our language. Hearing loss can affect either one or both ears. When one ear is affected it causes problems with directional hearing. Directional hearing provides the ability to determine from which direction a sound came. Lacking this ability can cause confusion within individuals who have hearing loss in one ear.

Temporary and Permanent

- PTS (Permanent Threshold Shift): the part of the hearing loss subsequent to an acoustic trauma that will never be recovered. PTS is measured in decibels.
- TTS (Temporary Threshold Shift): the hearing loss that will be recovered after a couple of days. Also called auditory fatigue. TTS is also measured in decibels.

TTS imperceptibly gives way to PTS.

In addition to hearing loss, other external symptoms of an acoustic trauma can be:

- Tinnitus
- Some pain in the ear
- Hyperacusis
- Dizziness or vertigo; in the case of vestibular damages, in the inner-ear

Tinnitus

An estimated 50 million Americans have some degree of tinnitus in one or both ears; 16 million of them have symptoms serious enough for them to see a doctor or hearing specialist. As many as 2 million become so debilitated by the unrelenting ringing, hissing, chirping, clicking, whooshing or screeching, that they cannot carry out normal daily activities.

Tinnitus is the largest single category for disability claims in the military, with hearing loss a close second. The third largest category is post-traumatic stress disorder, which itself may be accompanied by tinnitus and may exacerbate it.

Physiological Response

The symptoms mentioned above are the external signs of the physiological response to cochlear overstimulation. Here are some elements of this response:

- Broken and "bent" hairs of the hair-cells; damaged hair-cells degeneration. In humans, dead hair-cells are never replaced; the resulting hearing loss is thus irreparable.
- Inflammation of the exposed areas. This inflammation causes a bad blood flow in the exposed blood vessels (*vascular stasis*), and a bad oxygen supply for the liquid inside the cochlea (*endolymphatic hypoxia*) Those noxious conditions worsen the damaged hair-cells degeneration.
- Synaptic damages, by excitotoxicity. Noise overstimulation causes an excessive release of glutamate, causing the postsynaptic bouton to swell and burst. However the neuron connection can be repaired, and the hearing loss only caused by the "wiring" (i.e. excitotoxicity) can thus be recovered within 2–3 days.

Quality of Life

NIHL has implications on quality of life that extend beyond related symptoms and the ability to hear. The annual disability-adjusted life years (DALYs) were estimated for noise-exposed U.S. workers. DALYs represent the number of healthy years lost due to a disease or other health condition. They were defined by the 2013 Global Burden of Disease (GBD) Study. The DALYs calculation accounts for life limitations experienced because of hearing loss as a lost portion of a healthy year of life. So the results indicate the number of healthy years lost by a group of people over a specific time period.

NIOSH used DALYs to estimate the impact of hearing loss on quality of life in the CDC *Morbidity and Mortality Weekly Report* article "Hearing Impairment Among Noise-Exposed Workers in the United States, 2003-2012." It reported that 2.5 healthy years were lost each year for every 1,000 noise-exposed U.S. workers because of hearing impairment (hearing loss that impacts day-to-day activities). These lost years were shared among the 13% of workers with hearing impairment (about 130 workers out of each 1,000 workers). Mining, Construction and Manufacturing workers lost more healthy years than workers in other industry sectors; specifically and respectively in those sectors, 3.5, 3.1 and 2.7 healthy years were lost each year for every 1,000 workers.

Negative Impacts

The negative impacts of NIHL on one's ability to reciprocate communication, socialize and interact with society are largely invisible. Hearing loss, in general, is not just an issue of volume; individuals may experience difficulty in understanding what is said over the phone, when several people are talking at once, in a large space, or when the speaker's face cannot be seen. Subsequently, challenging social interactions can negatively lead to decreased self-esteem, shame, and fear. This can be more acutely felt by those who experience hearing impairment or loss earlier in life, rather than later when it is more socially accepted. Such psychosocial states, regardless of age, can lead to social isolation, which is known to negatively impact one's overall health and well-being. The compounding impacts can also lead to depression, especially if hearing impairment leads to tinnitus. Research suggests that those with hearing impairment or loss may be at a greater risk for deterioration of quality of life, as captured by a quote from Helen Keller: "Blindness cuts us off from things, but deafness cuts us off from people." Hearing impairment and loss of hearing, regardless of source or age, also limits experiencing the many benefits of sound on quality of life. In addition to the interpersonal social benefits, new studies suggest the effects of nature sounds, such as birds chirping and water, can positively affect an individual's capacity to recover after being stressed or to increase cognitive focus.

Quality of Life Questionnaire

Hearing loss is typically quantified by results from an audiogram; however, the degree of loss of hearing does not predict the impact on one's quality of life. The impact that NIHL can have on daily life and psychosocial function can be assessed and quantified using a validated questionnaire tool, such as the Hearing Handicap Inventory for the Elderly (HHIE). The HHIE is considered a "useful tool for quantifying the perceived emotional and social/situational consequences of hearing loss." The original tool was designed to test adults 65 years of age and older; however, modified versions exist. For adults the Hearing Handicap Inventory for Adults (HHIA) can be used and for adolescents the modified 28-item Hearing Environments And Reflection on Quality of Life (HEAR-QL-28) can be used. The HHIA, for example, is a 25-item questionnaire that asks both social and emotional-specific questions such as: Does a hearing problem cause you to avoid groups of people?" (social) and "Does a hearing problem cause you to feel frustrated when talking to members of your family?" (emotional). Response options are yes, no and sometimes.

Cause

The ear can be exposed to short periods of sound in excess of 120 dB without permanent harm — albeit with discomfort and possibly pain; but long term exposure to sound levels over 80 dB can cause permanent hearing loss.

There are two basic types of NIHL:

- NIHL caused by acoustic trauma and
- gradually developing NIHL.

Acute Acoustic Trauma

NIHL caused by acute acoustic trauma refers to permanent cochlear damage from a one-time exposure to excessive sound pressure. This form of NIHL commonly results from exposure to high-intensity sounds such as explosions, gunfire, a large drum hit loudly, and firecrackers.

Perceived Harmfulness vs. Actual Harmfulness

The discomfort threshold is the loudness level from which a sound starts to be felt as "too loud" and thus painful by an individual. Industry workers tend to have a higher discomfort threshold (i.e. the sounds must be louder to feel painful than for non-industry workers), but the sound is just as harmful to their ears. Industry workers often suffer from NIHL because the discomfort threshold is not a relevant indicator of the harmfulness of a sound.

Gradually Developing

Gradually developing NIHL refers to permanent cochlear damage from repeated exposure to loud sounds over a period of time. Unlike acoustic trauma, this form of NIHL does not occur from a single exposure to a high-intensity sound pressure level. Gradually developing NIHL can be caused by multiple exposures to excessive noise in the workplace or any source of repetitive, frequent exposures to sounds of excessive volume, such as home and vehicle stereos, concerts, nightclubs, and personal media players.

Workplace

About 22 million workers are exposed to hazardous noise, with additional millions exposed to solvents and metals that could put them at increased risk for hearing loss. Occupational hearing loss is one of the most common occupational diseases. 49% of male miners have hearing loss by the age of 50. By the age of 60, this number goes up to 70%. Construction workers also suffer an elevated risk. A screening program focused on construction workers employed at US Department of Energy facilities found 58% with significant abnormal hearing loss due to noise exposures at work. Occupational hearing loss is present in up to 33% of workers overall. Occupational exposure to noise causes 16% of adult disabling hearing loss worldwide.

The following is a list of occupations that are most susceptible to hearing loss:

- Agriculture
- Mining
- Construction
- Manufacturing

- Utilities
- Transportation
- Military
- Musicians
- Orchestra conductors

Among Musicians

Musicians, from classical orchestras to rock groups, are exposed to high decibel ranges. Some rock musicians experience noise-induced hearing loss from their music, and some studies have found that "symphonic musicians suffer from hearing impairment and that the impairment might be ascribed to symphonic music."

Music-induced hearing loss is still a controversial topic for hearing researchers. While some populational studies have shown that the risk for hearing loss increases as music exposure increases, other studies found little to no correlation between the two. Experts at the 2006 "Noise-Induced Hearing Loss in Children at Work and Play" Conference agreed that further research into this field was still required before making a broad generalization about music-induced hearing loss.

Given the extensive research suggesting that industrial noise exposure can cause sensorineural hearing loss a link between hearing loss and music exposures of similar level and duration (to industrial noise) seems highly plausible. Determining which individuals or groups are at risk for such exposures may be a difficult task. Recent research suggests that despite concerns about the proliferation of personal music players, in fact discos, concerts and live music events may be more hazardous to youth's hearing. People from ages to 6–19 have an approximately 15% rate of hearing loss. Recommendations for musicians to protect their hearing have been released in 2015.

Workplace Standards

In the United States, the Occupational Safety and Health Administration (OSHA) describes standards for occupational noise exposure. OSHA states that an employer must implement hearing conservation programs for employees if the noise level of the workplace is equal to or above 85 dB(A) for an averaged eight-hour time period. OSHA also states that "exposure to impulsive or impact noise should not exceed 140 dB peak sound pressure level". The National Institute for Occupational Safety and Health (NIOSH) recommends that all worker exposures to noise should be controlled below a level equivalent to 85 dBA for eight hours to minimize occupational noise induced hearing loss. NIOSH also recommends a 3 dBA exchange rate so that every increase by 3 dBA doubles the amount of the noise and halves the recommended amount of exposure time. The United States Department of Defense (DoD) instruction 605512 has some differences from OSHA 1910.95 standard, for example, OSHA 1910.95 uses a 5 dB exchange rate and DoD instruction 605512 uses a 3 dB exchange rate.

There are programs that seek to increase compliance and therefore effectiveness of hearing protection rules; the programs include the use of hearing tests and educating people that loud sound is dangerous

Employees are required to wear hearing protection when it is identified that their eight-hour time weighted average (TWA) is above the exposure action value of 90 dB. If subsequent monitoring shows that 85 dB is not surpassed for an eight-hour TWA, the employee is no longer required to wear hearing protection.

In the European Union, directive 2003/10/EC mandates that employers shall provide hearing protection at noise levels exceeding 80 dB(A), and that hearing protection is mandatory for noise levels exceeding 85 dB(A). Both values are based on 8 hours per day, with a 3 dB exchange rate.

A 2012 Cochrane review found low-quality evidence that legislation to reduce noise in the workplace was successful in reducing exposure both immediately and long-term.

Mechanisms

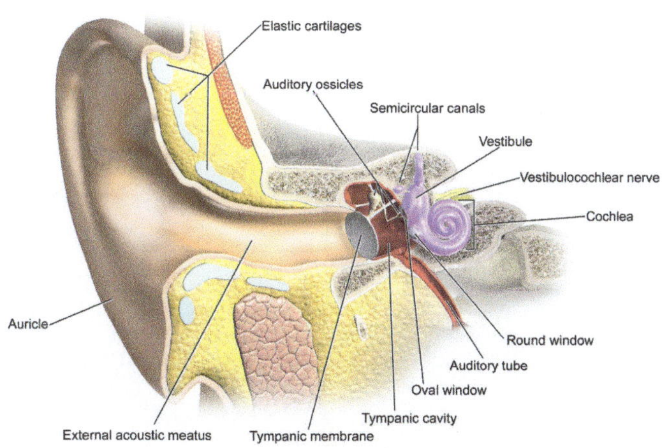

The Anatomy of the Ear

The outer ear receives sound, transmitted through the ossicles of the middle ear to the inner ear, where it is converted to a nervous signal in the cochlear and transmitted along the vestibulocochlear nerve

NIHL occurs when too much sound intensity is transmitted into and through the auditory system. An acoustic signal from a sound source, such as a radio, enters into the external auditory canal, and is funneled through to the tympanic membrane (eardrum). The tympanic membrane acts as an elastic diaphragm driving the ossicular chain of the middle ear system into motion. The middle ear ossicles transfer mechanical energy to the cochlea by way of the stapes footplate hammering against the oval window of the cochlea. This hammering causes the fluid within the cochlea (perilymph and endolymph) to push against the stereocilia of the hair cells, which then transmit a signal to the central auditory system within the brain. Different groups of hair cells are responsive to and conductive of different frequencies. There are two known biological mechanisms of NIHL from excessive sound intensity: damage to the hair cells and damage to the myelination or synaptic regions of auditory nerves.

Hair Cell Damage or Death

When the ear is exposed to excessive sound levels or loud sounds over time, the overstimulation of the hair cells leads to heavy production of reactive oxygen species, leading to oxidative cell death. In animal experiments, antioxidant vitamins have been found to reduce hearing loss even when administered the day after noise exposure. They were not able to fully prevent it. Damage ranges from exhaustion of the "hair" (hearing) cells in the ear to loss of those cells NIHL is therefore the consequence of overstimulation of the hair cells and supporting structures. Structural damage to hair cells (primarily the outer hair cells) will result in hearing loss that can be characterized by an attenuation and distortion of incoming auditory stimuli.

During cell death 'scars' develop, which prevent potassium rich fluid on the endolymph from mixing with the fluid on the basal domain. The fluids are kept from mixing because the potassium rich fluid is toxic to the neuronal endings and can damage hearing of the entire ear. If the endolymph fluid mixes with the fluid on the basal domain the neurons become depolarized, causing complete hearing loss. In addition to complete hearing loss, if the area is not sealed and leakage continues further tissue damage will occur. The 'scars' that form to replace the damaged hair cell are caused by supporting hair cells undergoing apoptosis and sealing the reticular lamina, which prevents fluid leakage. The cell death of two supporting hair cells rapidly expands their apical domain, which compresses the hair cell beneath its apical domain.

Nerve Damage

Recent studies have investigated additional mechanisms of NIHL involving delayed or disabled electrochemical transmission of nerve impulses from the hair cell to and along the auditory nerve. In cases of extreme acute acoustic trauma, a portion of the postsynaptic dendrite (where the hair cell transfers electrochemical signal to the auditory nerve) can rupture from overstimulation, temporarily stopping all transmission of auditory input to the auditory nerve. This is known as excitotoxicity. Usually, this sort of rupture heals within about five days, resulting in functional recovery of that synapse. While healing, an over-expression of glutamate receptors can result in temporary tinnitus, or ringing in the ears. Repeated ruptures at the same synapse may eventually fail to heal, leading to permanent hearing loss.

Acoustic over-exposure can also result in decreased myelination at specific points on the auditory nerve. Myelin, an insulating sheath surrounding nerve axons, expedites electrical impulses along nerves throughout the nervous system. Thinning of the myelin sheath on the auditory nerve significantly slows the transmission of electrical signals from hair cell to auditory cortex, reducing comprehension of auditory stimuli by delaying auditory perception, particularly in noisy environments.

Individual Susceptibility Towards Noise

There appear to be large differences in individual susceptibility to NIHL. The following factors have been implicated:

- missing acoustic reflex
- previous sensorineural hearing loss
- a bad general health state: bad cardiovascular function, insufficient intake of oxygen, a high platelet aggregation rate; and most importantly, a high viscosity of the blood
- cigarette smoking
- exposure to ototoxic chemicals (medication or environmental chemicals that can damage the ear), including certain solvents and heavy metals
- type 2 diabetes

Diagnosis

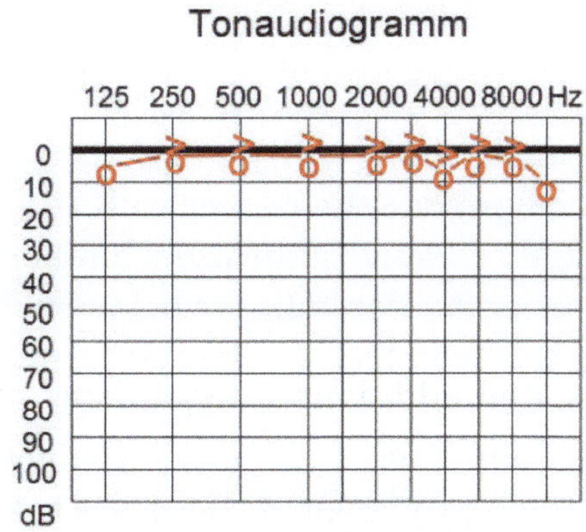

Example of an audiogram.

Both NIHL caused by acoustic trauma and gradually-developed-NIHL can often be characterized by a specific pattern presented in audiological findings. NIHL is generally observed to affect a person's hearing sensitivity in the higher frequencies, especially at 4000 Hz. "Noise-induced impairments are usually associated with a notch-shaped high-frequency sensorineural loss that is worst at 4000 Hz, although the notch often occurs at 3000 or 6000 Hz, as well". Doctoral students at the University of Iowa have termed this notch, specific to a noise-induced etiology, a "muna." The symptoms of NIHL are usually presented equally in both ears.

This typical 4000 Hz notch is due to the transfer function of the ear. Indeed, as any object facing a sound, the ear acts as a passive filter (-although the inner ear is not an absolute passive filter, as the outer hair cells provide active mechanisms). A passive filter is a low pass : the high frequencies are more absorbed by the object, as high frequencies impose a higher pace of compression-decompression to the object. Thus, the high frequency harmonics of a sound are more harmful to the inner-ear.

However, not all audiological results from people with NIHL match this typical notch. Often a decline in hearing sensitivity will occur at frequencies other than at the typical 3000–6000 Hz range. Variations arise from differences in people's ear canal resonance, the frequency of the harmful acoustic signal, and the length of exposure. As harmful noise exposure continues, the commonly affected frequencies will broaden and worsen in severity. "NIHL usually occurs initially at high frequencies (3, 4, or 6 kHz), and then spreads to the low frequencies (0.5, 1, or 2 kHz)".

Prevention

NIHL can be prevented through the use of simple, widely available, and economical tools. This includes but is not limited to personal noise reduction through the use of ear protection (i.e. earplugs and earmuffs), education, and hearing conservation programs. For the average person, there are three basic things that can be kept in mind to reduce NIHL, "walk away, turn it down, protect your ears."

Personal Noise Reduction Devices

Personal noise reduction devices can be passive, active or a combination. Passive ear protection includes earplugs or earmuffs which can block noise up to a specific frequency. Earplugs and earmuffs can provide the wearer with 10 dB to 40 dB of attenuation. However, use of earplugs is only effective if the users have been educated and use them properly; without proper use, protection falls far below manufacturer ratings. Higher consistency of performance has been found with custom-molded earplugs. Because of their ease of use without education, and ease of application or removal, earmuffs have more consistency with both compliance and noise attenuation. Active ear protection (electronic pass-through hearing protection devices or EPHPs) electronically filter out noises of specific frequencies or decibels while allowing the remaining noise to pass through.

Education

Education is key to prevention. Before hearing protective actions will take place, a person must understand they are at risk for NIHL and know their options for prevention.

Hearing protection programs have been hindered by people not wearing the protection for various reasons, including the desire to converse, uncomfortable devices, lack of concern about the need for protection, and social pressure against wearing protection. Although youth are at risk for hearing loss, one study found that 96.3% of parents did not believe their adolescents were at risk, and only 69% had talked to their children about hearing protection; those aware of NIHL risks were more likely to talk to their teens.

A systematic review of the effectiveness of interventions to promote the use of hearing protection devices such as earplugs and earmuffs among workers found that tailored interventions improve the average use of such devices when compared with no intervention. Tailored interventions involve the use of communication or other types of interventions that are specific to an individual or a group and aim to change behavior. Mixed interventions such as mailings, distribution of hearing protection devices, noise assessments, and hearing testing are also more effective in improving the use of hearing protection devices compared with hearing testing alone. Programs that increased the proportion of workers wearing hearing protection equipment did reduce overall hearing loss. However, a 2012 Cochrane review found that hearing loss prevention programs have not shown as much success with educational messages as with use of hearing protective devices; better implementation and reinforcement of these programs are needed.

Personal Listening Devices

A 2012 study looked at the safety of using personal listening devices as well as the comparative safety of earbuds, in-ear earphones, and supra-aural earphones (over the ear). The study found that since supra-aural earphones blocked out background noise better than the other earphones, sound was listened to at a quieter, and safer volume. Ear buds blocked the least amount of environmental noise and seemed to result in the most noise pressure. Recommendations for safe use of personal listening devices included: keeping the volume at the lowest possible setting, taking care to keep volume below 50% even in loud settings such as buses or airplanes, and selecting an earphone that is better at attenuating environmental noise.

Hearing Conservation Programs

Workers in general industry who are exposed to noise levels above 85 dBA are required by the Occupational Safety and Health Administration (OSHA) to be in a hearing conservation program (HCP), which includes noise measurement, noise control, periodic audiometric testing, hearing protection, worker education, and record keeping. Twenty-four states, Puerto Rico, and the U.S. Virgin Islands have OSHA-approved state plans and have adopted their own standards and enforcement policies. Most of these state standards are identical to those of federal OSHA. However, some states have adopted different standards or may have different enforcement policies. Most health and safety regulations are designed to keep damage risk within "acceptable limits" — that is, some people are likely to incur a hearing loss even when exposed to less than the maximum daily amount of noise specified in a regulation. Hearing conservation programs in other arenas (schools, military) have become more common, and it has been established that unsafe listening behaviors, such as listening to loud noise for extended periods of time without protection, persist despite knowledge of potential hearing loss effects.

However, it is understood that HCPs are designed to change behavior, which is known to be a complex issue that requires a multi-faceted approach. According to Keppler et al. in their 2015 study of such programming, they cite the necessary attitude change towards the susceptibility of risk and degree of severity of hearing loss. Among young adults, the concept of severity is most crucial because it has been found that behavior change may not occur unless an individual experiences NIHL or similarly related NIHL tinnitus, furthering warranting a multi-pronged approach based on hearing conservation programming and education.

Other Initiatives

There are a variety of public awareness programs as well as available curricula to teach awareness messages becoming available. One such program is Dangerous Decibels, whose mission is to "significantly reduce the prevalence of noise induced hearing loss and tinnitus through exhibits, education and research. We're hEAR for You is a small non-profit that distributes information and ear plugs at concert and music festival venues. The Buy Quiet program was created to combat occupational noise exposures by promoting the purchase of quieter tools and equipment and encourage manufacturers to design quieter equipment. The National Institute for Occupational Safety and Health partnered with the National Hearing Conservation Association in 2007 to establish the Safe-in-Sound Excellence and Innovation in Hearing Loss Prevention Awards to recognize organizations that are successfully implementing hearing loss prevention concepts into their daily routines.

Pharmacalogical

There is evidence that hearing loss can be minimized by taking high doses of magnesium for a few days, starting as soon as possible after exposure to the loud noise. A magnesium-high diet also seems to be helpful as an NIHL-preventative if taken in advance of exposure to loud noises. Consuming sizable amounts of magnesium can be potentially harmful, so this treatment should be followed with caution.

Sound or Stress Training

Despite different people having different thresholds for what noises are painful, this pain threshold had no correlation with which noises cause hearing damage. The ear can not get more resistant to noise harmfulness by training it to noise. The cochlea is partially protected by the acoustic reflex, but being frequently exposed to noise does not lower the reflex threshold. It had been observed that noise conditioning (i.e. exposure to loud non-traumatizing noise) several hours prior to the exposure to traumatizing sound level, significantly reduced the damages inflicted to the hair-cells. The same "protective effect" was also observed with other stressors such as heat-shock conditioning and stress (by restraint) conditioning. This "protective effect" only happens if the traumatizing noise is presented within an optimum interval of time after the sound-conditioning session (-24 hours for a 15 min. sound-conditioning; no more protection after 48 hours).

This "protective effect" had long been thought to involve the active mechanisms of the outer hair cells and the efferent system commanding them. The contractile effect of the outer hair cells, activated by the efferent nervous system has been proven to provide a protective effect against acoustic trauma.

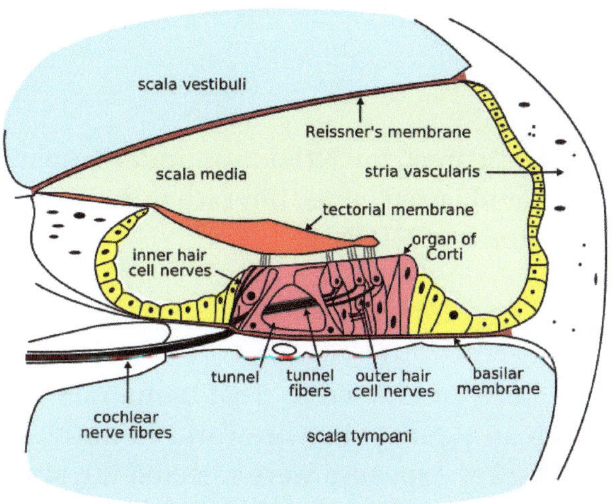

Cross-section of the cochlea. The inner hair cells are connected to **afferent** nerve fibers, and the outer hair cells are connected to **efferent** nerve fibers.

However, a 2006 study revealed a different protective mechanism for stress conditioning. The study revealed that the stressor (sound, heat, or stress) conditioning increases the receptibility to glucocorticoid, a kind of anti-inflammatory hormone. The effects of glucocorticoid thus mitigate the inflammation from an acoustic trauma that can lead to hearing loss. In fact, high doses of corticoids are often prescribed by physicians after an acoustic-trauma in order to mitigate the inflammatory response.

Summarized, sound (or other stressor) conditioning is a pre-emptive medication against cochlea inflammation. It does not make the ear more resistant to noise. It reduces the inflammation caused by the acoustic trauma, which would cause subsequent damages to hair cells. While an anti-inflammatory medication would increase the quantity of anti-inflammatory hormone in the whole body, noise conditioning increases the number of receptors for the anti-inflammatory hormone, and only in the areas where it is much needed (i.e. cochlea).

Physiological Response

- stressor (noise, heat shock or stress) conditioning activates hormonal glands: the HPA axis. Note that the HPA axis is associated to the immune system
- this HPA axis activation results in the up regulation of glucocorticoid receptors (GR) in the cochlea and the paraventricular nucleus (PVN) of the hypothalamus. Note that the glucocorticoid hormone is a kind of immune-reaction-inhibitor, including the inflammation reaction.
- This up regulation of GR thus prevents GR down regulation induced by acoustic trauma
- The protective effect of noise-conditioning is blocked by adrenalectomy or pharmacological treatment with RU486+ metyrapone (a glucocorticoid receptor antagonist).

Treatment

Treatment options that offer "cures" for NIHL are under research and development. Currently there are no commonly used cures, but rather assistive devices and therapies to try and manage the symptoms of NIHL.

Acute Acoustic Trauma

Several clinical trials have been conducted to treat temporary NIHL occurring after a traumatic noise event, such as a gunshot or firework. In 2007, individuals with acute acoustic trauma after firecracker exposure were injected intratympanically with a cell permeable ligand, AM-111. The trial found AM-111 to have a therapeutic effect on at least 2 cases of those with acute trauma. Treatment with a combination of prednisolone and piracetam appeared to rescue patients with acute trauma after exposure to gunshots. However, those who received the treatment within an hour of exposure had higher rates of recovery and significantly lower threshold shifts compared to those who received treatment after 1 hour.

Additionally, clinical trials using antioxidants after a traumatic noise event to reduce reactive oxygen species have displayed promising results. Antibiotic injections with allopurinol, lazaroids, α-D-tocopherol, and mannitol were found to reduce the threshold shift after noise exposure. Another antioxidant, Ebselen, has been shown to have promising results for both TTS and PTS. Ebselen mimics gluthathione peroxide, an enzyme that has many functions, including scavenging hydrogen peroxide and reactive oxygen species. After noise exposure, gluthathione peroxide decreases in the ear. An oral administration of ebselen in both preclinical tests on guinea pigs and human trials indicate that noise induced TTS and PTS was reduced.

Gradually Occurring NIHL

At the present time, no established clinical treatments exist to reverse the effects of permanent NIHL. However, current research for the possible use of drug and genetic therapies look hopeful. In addition, management options such as hearing aids and counseling exist.

Many studies have been conducted looking at regeneration of hair cells in the inner ear. While hair cells are generally not replaced through cell regeneration, mechanisms are being studied to induce replacement of these important cells. One study involves the replacement of damaged hair cells with regenerated cells, via the mechanism of gene transfer of atonal gene Math1 to pluripotent stem cells within the inner ear. Other atonal genes are being studied to induce regeneration of hair cells in the inner ear.

Management

For people living with NIHL, there are several management options that can improve the ability to communicate including counseling and the use of hearing aids and FM systems. The prognosis has improved with the recent advancements in digital hearing aid technology, such as directional microphones, open-fit hearing aids, and more advanced algorithms. Annual audiological evaluations are recommended to monitor any changes in a patient's hearing and to modify hearing-aid prescriptions.

A systematic-review conducted by the American Academy of Audiology Task Force On the Health-Related Quality of Life Benefits of Amplification in Adults found the use of hearing aids to increase quality of life. The review pertained to adults who experienced sensorineural hearing loss, which can be caused by excessive, loud noise.

Epidemiology

Currently, the World Health Organization estimates that nearly 360 million people suffer from moderate to profound hearing loss from all causes. Incidence of hearing loss has traditionally been attributed to occupational or firearm-related exposure, however, more recent research suggests that the trend is shifting as prevalence rates increase amongst children and adolescents is increasing. According to a 2015 review, "Hearing loss due to recreational exposure to loud sounds" by the WHO World Health Organization (WHO), an estimated 1.1 billion young people may be at risk for hearing loss caused by unsafe listening practices. The over-exposure to excessive loud noise is attributed to recreational exposure, such as the use of personal audio devices with music at high volumes for long durations raises, or social settings such as bars, entertainment and sporting events.

A review of data from the National Health and Nutrition Examination Survey in the United States found that the prevalence of high frequency hearing loss, which can be noise-induced, among adolescent 12–19 years old increased significantly from 3.5% to

5.3% from 1994 until 2006, which equates to 1 in 20 children in the U.S. Although research is still limited, it suggests that increased exposure to loud noise through personal listening devices is a risk factor to this type of acquired hearing loss. The ubiquitous use of smartphones across the world is increasing at a rate "faster than our ability to assess their potential health consequences." Despite limited research, it is documented that 90% young adults across U.S. and European surveys have reported use of personal devices for several hours a day at maximum volume. At times, this sound level can be likened to a jet engine, exceeding 120 decibels. Recent research suggests stronger correlations found between extended duration and/or elevated usage of personal listening devices and hearing loss.

Non-occupational noise exposure is not regulated or governed in the same manner as occupational noise exposure; therefore prevention efforts rely heavily on education awareness campaigns and public policy. The WHO cites that nearly half of those affected by hearing loss could have been prevented through primary prevention efforts such as:

"reducing exposure (both occupational and recreational) to loud sounds by raising awareness about the risks; developing and enforcing relevant legislation; and encouraging individuals to use personal protective devices such as earplugs and noise-cancelling earphones and headphones."

Occupational Hearing Loss

Occupational hearing loss (OHL) is hearing loss that occurs as a result of occupational hazards. OHL, damage to one or both ears from exposures related to one's occupation, is a large but preventable problem. Organizations such as the Occupational Safety and Health Administration (OSHA) and the National Institute for Occupational Safety and Health (NIOSH) work with employers and workers to reduce or eliminate completely hazards to hearing. Occupational hearing loss is one of the most common work-related illness in the United States.

Hazards to the hearing of workers include industrial noise, and exposure to various ototoxic chemicals. These exposures may also interact to cause more damage than either one would by itself. Many chemicals have not been tested for ototoxicity, so unknown threats may exist.

A 2016 study by NIOSH found that the mining sector had the highest prevalence of hearing impairment at 17%, followed by the construction sector (16%) and the manufacturing sector (14%). The public safety sector had the lowest rate of hearing impairment, at 7%.

Personal protective equipment, administrative controls, and engineering controls can all work to reduce exposure to noise and chemicals, either by providing the worker with protection such as earplugs, or by reducing the noise at the source or limiting the time or level of exposure.

Background

Occupational hearing loss is defined as damage to either or both ears, at the inner ear or auditory nerve, that results from an exposure in a person's occupation. Although high levels of noise are the main cause of occupational hearing loss there are also other factors in the work environment that can result in it. Chemicals, foreign bodies, vibration, barotrauma, along with other hazards can result in hearing loss. These losses that these workers obtain, affect many aspects of their life, mainly social interactions.

Within the United States of America, approximately 10 million people have noise-related hearing loss. Over twice that number are occupationally exposed to dangerous noise levels. Hearing loss accounted for a sizable percentage of occupational illness in 2007, at 14% of cases.

There are many organizations, including the Occupational Safety and Health Administration (OSHA) and the National Institute for Occupational Safety and Health (NIOSH) working to understand the causes of occupational hearing loss and how it can be prevented. They work to produce regulations and guidelines to help protect the hearing of workers in all occupations.

Causes

Noise Exposure

Exposure to noise can cause vibrations able to cause permanent damage to the ear. Both the volume of the noise and the duration of exposure can influence the likelihood of damage. Sound is measured in units called decibels, which is a logarithmic scale of sound levels that corresponds to the level of loudness that an individual's ear would perceive. Because it is a logarithmic scale, even small incremental increases in decibels correlate to large increases in loudness, and an increase in the risk of hearing loss.

Sounds above 80 dB have the potential to cause permanent hearing loss. The intensity of sound is considered too great and hazardous if someone must yell in order to be heard. Ringing in the ears upon leaving work is also indicative of noise that is at a dangerous level. Farming, machinery work, and construction are some of the many occupations that put workers at risk of hearing loss.

OSHA's current permissible exposure limit (PEL) for workers is an average of 90 dB over an 8-hour work day. They also use a 5 dB exchange rate, where an increase in 5 dB for a sound corresponds to the amount of time workers may be exposed to that particular source of sound being halved. For example, workers cannot be exposed to a sound level of 95 dB for more than 4 hours per day, or to sounds at 100 dB for more than 2 hours per day. Employers who expose workers to 85 dB or more for 8 hour shifts are required to provide hearing exams and protection, monitor noise levels, and provide training.

Sound level meters and dosimeters are two types of devices that are used to measure sound levels in the workplace. Dosimeters are typically worn by the employee to measure their own personal sound exposure. Other sound level meters can be used to double check dosimeter measurements, or used when dosimeters cannot be worn by the employees. They can also be used to evaluate engineering controls aimed at reducing noise levels.

Ototoxic Chemical Exposure

Chemically-induced hearing loss (CIHL) is a potential result of occupational exposures. Certain chemical compounds may have ototoxic effects. Exposure to organic solvents, heavy metals, and asphyxiants such as carbon monoxide can all cause hearing loss. These chemicals can be inhaled, ingested, or absorbed through the skin. Damage can occur to either the inner ear or the auditory nerve.

Both noise and chemical exposures are common in many industries, and can both contribute to hearing loss simultaneously. Damage may be more likely or more severe if both are present. Industries in which combinations of exposures may exist include construction, fiberglass, metal manufacturing, and many more.

Certain medications may also have the potential to cause hearing loss.

It is estimated that over 22 million workers are exposed to dangerous noise levels, and 10 million are exposed to solvents that could potentially cause hearing loss every year, with an unknown number exposed to other ototoxic chemicals.

Prevention

Several methods of prevention exists to eliminate or reduce the hearing loss caused by workplace exposure.

Hierarchy of Controls

The hierarchy of controls provides a visual guide to the effectiveness of the various workplace controls set in place to eliminate or reduce exposure to occupational hazards, including noise or ototoxic chemicals. The hierarchy includes the following from most effective to least effective:

- Elimination: complete removal of the hazard
- Substitution: replacement the offers a smaller risk
- Engineering controls: physical changes to reduce exposure

- Administrative controls: changes in work procedures or training
- Personal protective equipment (PPE): individual equipment to reduce exposure, e.g. earplugs

Engineering Controls

Engineering control is the highest in the hierarchy of risk reduction methods when elimination of the hazard is not possible. These types of controls typically involve making changes in equipment or other changes to minimize the level of noise that reaches a worker's ear. They may also involve measures such as barriers between the worker and the source of the noise, mufflers, regular maintenance of the machinery, or substituting quieter equipment. Examples of noise control strategies adopted in the workplace can be seen among the winners of the Safe-in-Sound Excellence in Hearing Loss Prevention Awards.

Administrative Controls

Administrative control, behind engineering control, is the next best form of prevention of noise exposure. They can either reduce the exposure to noise, or reduce the decibel level of the noise itself. Limiting the amount of time a worker is allowed to be around an unsafe level of noise exposure, and creating procedures for operation of equipment that could produce harmful levels of noise are both examples of administrative controls.

Personal Protection

Elimination or reduction of the source of noise or chemical exposure is ideal, but when that is not possible or adequate, wearing personal protective equipment (PPE) such as earplugs or earmuffs can help reduce the risk of hearing loss due to noise exposure. PPE should be a last resort and not be used in substitution for engineering or administrative controls. It is important that workers are properly trained on the use of PPE to ensure proper protection.

Other Initiatives

In addition to the hierarchy of controls, other programs have been created to promote the prevention of hearing loss in the workplace. For example, the Buy Quiet program wad created to encourage the purchase of quieter tools and machinery in the workplace. Additionally, the Safe-n-Sound Award was created to recognize organizations that excel in preventing occupational hearing loss.

History

Occupational hearing loss is a very present industrial issue that has been noticed since the Industrial Revolution. As industrial society continues to grow, this issue is becoming increasingly detrimental due to the exposure of chemicals and physical objects. Millions of employees have been affected by occupational hearing loss, especially in industry. Industrialized countries see most of these damages as they result in both economic and living problems.

Within the United States of America alone, 10 of the 28 million people that have experienced hearing loss related to noise exposure. Rarely do workers express concerns or complaints regarding Occupational hearing loss. In order to gather relevant information, workers who have experienced loud work environments are questioned regarding their hearing abilities during everyday activities. When analyzing OHP, it is necessary to consider family history, hobbies, recreational activities, and how they could play a role in a person's hearing loss. In order to test hearing loss, audiometers are used to and are adjusted to American National Standards Institute (ANSI) regulations. The Occupation and Safety Health Association (OSHA) of the United States of America requires a program that conserves hearing when noise levels are greater than 85 dB. This program includes: 1."Monitoring to assess and record noise levels." 2. "Periodic audiometry." 3. "Noise Control" 4. "Education and record keeping." Occupational hearing loss is very preventable.

Health Effects from Noise

Traffic is the main source of noise pollution in cities.

Noise health effects are the health consequences of regular exposure, to consistent elevated sound levels. Elevated workplace or environmental noise can cause hearing impairment, hypertension, ischemic heart disease, annoyance, and sleep disturbance. Changes in the immune system and birth defects have been also attributed to noise exposure.

Although some presbycusis may occur naturally with age, in many developed nations the cumulative impact of noise is sufficient to impair the hearing of a large fraction of the population over the course of a lifetime. Noise exposure also has been known to induce tinnitus, hypertension, vasoconstriction, and other cardiovascular adverse effects.

Beyond these effects, elevated noise levels can create stress, increase workplace accident rates, and stimulate aggression and other anti-social behaviors. The most significant causes are vehicle and aircraft noise, prolonged exposure to loud music, and industrial noise. In Norway, road traffic has been demonstrated to cause almost 80% of the noise annoyances reported.

There may be psychological definitions of noise as well. Firecrackers may upset domestic and wild animals or noise-traumatized individuals. The most common noise-traumatized persons are those exposed to military conflicts, but often loud groups of people can trigger complaints and other behaviors about noise. Infants are easily startled by noise.

The social costs of traffic noise in EU22 are more than €40 billion per year, and passenger cars and lorries (trucks) are responsible for bulk of costs. Traffic noise alone is harming the health of almost every third person in the WHO European Region. One in five Europeans is regularly exposed to sound levels at night that could significantly damage health.

Noise also is a threat to other species in marine and terrestrial ecosystems alike.

Hearing Loss

The mechanism of hearing loss can be attributed to aging, infection, surgery, prolonged use of some medications, trauma, and to stereocilia of the cochlea, the principal fluid filled structure of the inner ear. The pinna combined with the middle ear amplifies sound pressure levels by a factor of twenty, so that extremely high sound pressure levels arrive in the cochlea, even from moderate atmospheric sound stimuli. Underlying pathology to the cochlea are reactive oxygen species, which play a significant role in noise-induced necrosis and apoptosis of the stereocilia.

Exposure to high levels of noise have differing effects within a given population, and the involvement of reactive oxygen species suggests possible avenues to treat or prevent damage to hearing and related cellular structures.

The elevated sound levels cause trauma to cochlear structure in the inner ear, which gives rise to irreversible hearing loss. A very loud sound in a particular frequency range can damage the cochlea's hair cells that respond to that range, thereby reducing the ear's ability to hear those frequencies in the future, however, loud noise in *any* frequency range has deleterious effects across the entire range of human hearing. The outer ear (visible portion of the human ear) combined with the middle ear amplifies sound levels by a factor of 20 when sound reaches the inner ear.

Age-related (Presbycusis)

Hearing loss is somewhat inevitable with age. Though older males exposed to significant occupational noise demonstrate significantly reduced hearing sensitivity compared to non-exposed peers. Differences in hearing sensitivity decrease with time and the two groups are indistinguishable by age 79. Over time, the detection of high-pitched sound frequencies becomes more difficult. This affects speech perception, particularly of those words involving sibilants and fricatives. Both ears tend to be affected.

Women exposed to occupational noise do not differ from their peers in hearing sensitivity, although they do hear better than their non-exposed male counterparts. Consistent exposure to loud music, young people in the United States have a rate of impaired hearing 2.5 times greater than their parents and grandparents. An estimated 50 million individuals will have in 2050.

Occupational Hearing Loss

Hearing loss as a result of occupational exposure is one of the most common work-related illnesses. In fact, every year in the United States, more than 22 million workers are exposed to hazardous noise and an additional 9 million are exposed to ototoxic chemicals. Those in certain occupations are at a higher risk of developing hearing loss as a result of the nature of their job. For example, musicians, miners, and those in manufacturing and construction may be exposed to higher and more constant noise levels.

Any workplace that exposes workers to excessive sound levels is required to ensure that the hearing function of their workers is adequately protected. The Occupational Safety and Health Administration (OSHA) provide laws and regulations clarified by the United States department of labor. Generated by the Occupational Safety and Health Act of 1970, OSHA was created with the mission to assure all working conditions are safe and healthy. All employers must follow this hearing amendment, including private sector employers, and excluding those who are self-employed, family farm workers, and government workers. Separate laws and regulations are enacted for specialized working environments. These include the Mine Safety and Health Administration (MSHA) and the Federal Railroad Administration (FRA), which provide specific regulation for miners and railroad workers.

The National Institute for Occupational Safety and Health (NIOSH) generated a criteria document consisting of a "best practice guide" recommended by OSHA to all employers. This document is recommended to be followed, but is not required as OSHA laws and regulation are. The regulations provided by NIOSH are much more conservative, and it provides scientific basis for occupational safety and health standards. For example, the permissible exposure limit for OSHA is >90 dBA, while NIOSH is >85 dBA. They recommend the implementation or promotion of these conservative standards alongside the standards of the designated administration (OSHA, MSHA, or FRA) of the employer.

OSHA, NIOSH, MSHA, and FRA were all created in response to an increase in noise induced hearing loss resulting from hazardous sound levels in industrial work environments. Noise can cause detrimental effects to the outer, middle, and inner ear, as well as the auditory nerve and central auditory system. In the outer ear, with high noise exposure the eardrum can rupture and bleed. In the middle ear, dislocation of the bones that reside in the middle ear (the ossicles) can occur. These problems occur when exposed to extremely high sound levels. In the inner ear, low, middle, and high sound levels can all cause damage to structures residing in the cochlea. This damage can cause what is called a 'noise notch' and hearing loss is evident at 3000 – 6000 Hz frequencies.

Cardiovascular Effects

Noise has been associated with important cardiovascular health problems. In 1999, the World Health Organization (WHO) concluded that the available evidence suggested a weak correlation between long-term noise exposure above 67-70 dB(A) and hypertension. More recent studies have suggested that noise levels of 50 dB(A) at night may also increase the risk of myocardial infarction by chronically elevating cortisol production.

Fairly typical roadway noise levels are sufficient to constrict arterial blood flow and lead to elevated blood pressure; in this case, it appears that a certain fraction of the population is more susceptible to vasoconstriction. This may result because annoyance from the sound causes elevated adrenaline levels trigger a narrowing of the blood vessels (vasoconstriction), or independently through medical stress reactions. Other effects of high noise levels are increased frequency of headaches, fatigue, stomach ulcers, and vertigo.

Stress

Research commissioned by Rockwool, a UK insulation manufacturer, reveals in the UK one third (33%) of victims of domestic disturbances claim loud parties have left them unable to sleep or made them stressed in the last two years. Around one in eleven (9%) of those affected by domestic disturbances claims it has left them continually disturbed and stressed. More than 1.8 million people claim noisy neighbours have made their life a misery and they cannot enjoy their own homes. The impact of noise on health is potentially a significant problem across the UK given that more than 17.5 million Britons (38%) have been disturbed by the inhabitants of neighbouring properties in the last two years. For almost one in ten (7%) Britons this is a regular occurrence.

The extent of the problem of noise pollution for public health is reinforced by figures collated by Rockwool from local authority responses to a Freedom of Information Act (FOI) request. This research reveals in the period April 2008 - 2009 UK councils received 315,838 complaints about noise pollution from private residences. This resulted in environmental health officers across the UK serving 8,069 noise abatement notices, or citations under the terms of the Anti-Social Behaviour (Scotland) Act.

Westminster City Council has received more complaints per head of population than any other district in the UK with 9,814 grievances about noise, which equates to 42.32 complaints per thousand residents. Eight of the top 10 councils ranked by complaints per 1,000 residents are located in London.

Annoyance

Because some stressful effects depend on qualities of the sound other than its absolute decibel value, the annoyance associated with sound may need to be considered in regard to health effects. For example, noise from airports or sudden Impulse noises are typically perceived as more bothersome than noise from traffic of equal volume. Annoyance effects of noise are minimally affected by demographics, but fear of the noise source and sensitivity to noise both strongly affect the 'annoyance' of a noise. Even sound levels as low as 40 dB(A) (about as loud as a refrigerator or library) can generate noise complaints and the lower threshold for noise producing sleep disturbance is 45 dB(A) or lower.

Other factors that affect the 'annoyance level' of sound include beliefs about noise prevention and the importance of the noise source, and annoyance at the cause (i.e. non-noise related factors) of the noise. For instance, in an office setting, audible telephone conversations and discussions between co-workers were considered to be irritating, depending upon the contents of the conversations. Many of the interpretations of the level of annoyance and the relationship between noise levels and resulting health symptoms could be influenced by the quality of interpersonal relationships at the workplace, as well as the stress level generated by the work itself. Evidence for impact on annoyance of long-term noise versus recent changes is equivocal.

Estimates of sound annoyance typically rely on weighting filters, which consider some sound frequencies to be more important than others based on their presumed audibility to humans. The older dB(A) weighting filter described above is used widely in the U.S., but underestimates the impact of frequencies around 6000 Hz and at very low frequencies. The newer ITU-R 468 noise weighting filter is used more widely in Europe. The propagation of sound varies between environments; for example, low frequencies typically carry over longer distances. Therefore, different filters, such as dB(B) and dB(C), may be recommended for specific situations.

Furthermore, studies have shown that neighborhood noise (consisting of noise from neighboring apartments, as well as noise within one's own apartment or home) can cause significant irritation and noise stress within people, due to the great deal of time people spend in their residences. This can result in an increased risk of depression and psychological disorders, migraines, and even emotional stress.

In the workplace, noise pollution is generally a problem once the noise level is greater than 55 dB(A). Selected studies show that approximately 35% to 40% of office workers find noise levels from 55 to 60 dB(A) extremely irritating. The noise standard in Germany for mentally stressful tasks is set at 55 dB(A), however, if the noise source is continuous, the threshold level for tolerability among office workers is lower than 55 dB(A).

One important effect of noise is to make a person's speech less easy to hear. The human brain compensates for background noise during speech production in a process called the Lombard effect in which speech becomes louder with more distinct syllables. However, this cannot fully remove the problems of communication intelligibility made in noise.

Infrasound (Low-frequency Sound)

Infrasound, sometimes referred to as low-frequency sound, is sound that is lower in frequency than 20 Hz (hertz) or cycles per second, the "normal" limit of human hearing. Hearing becomes gradually less sensitive as frequency decreases, so for humans to perceive infrasound, the sound pressure must be sufficiently high.

20 Hz is considered the normal low-frequency limit of human hearing. When pure sine waves are reproduced under ideal conditions and at very high volume, a human listener will be able to identify tones as low as 12 Hz.

One study has suggested that infrasound may cause feelings of awe or fear in humans. It also was suggested that since it is not consciously perceived, it may make people feel vaguely that odd or supernatural events are taking place.

Child Physical Development

The U.S. Environmental Protection Agency authored a pamphlet in 1978 that suggested a correlation between low-birthweight (using the World Health Organization definition of less than 2,500 g (~5.5 lb) and high sound levels, and also high rates of birth defects in places where expectant mothers are exposed to elevated sound levels, such as typical airport environs. Specific birth abnormalities included harelip, cleft palate, and defects in the spine.

According to Lester W. Sontag of The Fels Research Institute (as presented in the same EPA study): "There is ample evidence that environment has a role in shaping the physique, behavior, and function of animals, including man, from conception and not merely from birth. The fetus is capable of perceiving sounds and responding to them by motor activity and cardiac rate change." The effects of noise exposure are highest when it occurs between 15 and 60 days after conception, a period in which major internal organs and the central nervous system are formed.

Later developmental effects occur as vasoconstriction in the mother reduces blood flow and therefore oxygen and nutrition to the fetus. Low birth weights and noise were also associated with lower levels of certain hormones in the mother. These hormones are thought to affect fetal growth and to be good indicators of protein production. The difference between the hormone levels of pregnant mothers in noisy versus quiet areas increased as birth approached.

In a 2000 publication, a review of studies on birthweight and noise exposure note that while some older studies suggest that when women are exposed to >65 dB aircraft noise a small decrease in birthweight occurs, in a more recent study of 200 Taiwanese women including noise dosimetry measurements of individual noise exposure, the authors found no significant association between noise exposure and birth weight after adjusting for relevant confounders, e.g. social class, maternal weight gain during pregnancy, etc.

Cognitive Development

When young children are regularly exposed to levels of noise that interfere with speech, they may develop speech or reading difficulties, because auditory processing functions are compromised. Children continue to develop their speech perception abilities until they reach their teens. Evidence has shown that when children learn in noisier classrooms, they have more difficulties understanding speech than those who learn in quieter settings.

In a study conducted by Cornell University in 1993, children exposed to noise in learning environments experienced trouble with word discrimination, as well as various cognitive developmental delays. In particular, the writing learning impairment dysgraphia is commonly associated with environmental stressors in the classroom.

High noise levels have also been known to damage the physical health of small children. Children from noisy residences often have a heart rate that is significantly higher (by 2 beats/min on average) than those of children from quieter homes.

Regulations

Environmental noise regulations usually specify a maximum outdoor noise level of 60 to 65 dB(A), while occupational safety organizations recommend that the maximum exposure to noise is 40 hours per week at 85 to 90 dB(A). For every additional 3 dB(A), the maximum exposure time is reduced by a factor 2, e.g. 20 hours per week at 88 dB(A). Sometimes, a factor of two per additional 5 dB(A) is used, however, these occupational regulations are acknowledged by the health literature as inadequate to protect against hearing loss and other health effects. In an effort to prevent noise-induced hearing loss, many programs and initiative have been created, like the Buy Quiet program, which encourages employers to purchase quieter tools and equipment, and the Safe-In-Sound Award, which recognizes organizations with successful hearing loss prevention strategies.

With regard to indoor noise pollution in residences, the U.S. Environmental Protection Agency (EPA) has not set any restrictions on limits to the level of noise. Rather, it has provided a list of recommended levels in its *Model Community Noise Control Ordinance*, which was published in 1975. For instance, the recommended noise level for indoor residences is less than or equal to 45 dB.

Noise pollution control in residences is not funded by the federal government in part because of the disagreements in establishing causal links between sounds and health risks, since the effect of noise is often psychological and also, because it leaves no singular tangible trace of damage on the human body. For instance, hearing loss could be attributed to a variety of factors including age, rather than solely due to excessive exposure to noise. A state or local government is able to regulate indoor residential noise, however, such as when excessive noise from within a home causes disturbances to nearby residences.

In Canines

While people are often educated on the effects of noise exposure in humans, there are also different noise exposure effects in animals as well. An example of this would be in canines, and the noise exposure levels occurring within kennels. Canines experience this noise exposure whether it be a long stay at an animal shelter, or a weekend stay at a boarding facility.

Organizations like NIOSH and OSHA have different regulations when it comes to the noise exposure levels in industrial workers. Currently there are no regulations related to the noise exposure in canines even with such damaging effects related to their health. Health risks dogs are exposed to include ear damage and behavioral changes.

The average noise exposure in a kennel is greater than 100 dB SPL. According to OSHA these levels would yield in the use of hearing protection for the workers of those kennels due to the risk of noise induced hearing loss. The anatomical structures of the human and canine ear are very similar, so it is thought that these levels will negatively impact the hearing of canines in kennels. The ABR can be used to estimate the hearing threshold of canines, and can be used to show either a temporary threshold shift or permanent threshold shift after being exposed to excessive sound levels.

Behavioral effects to excessive noise exposure include hiding, urinating, defecating, panting, pacing, drooling, disregard to commands, trembling, and barking. These behavioral patterns pose a much greater problem to canines than meets the eye. All of these behavioral patterns are characteristics that result in a longer stay at the kennels before being adopted. A longer stay at the shelter results in a longer duration of noise exposure and therefore more likely to show either a temporary or permanent threshold shift in the canine's hearing.

These excessive noise levels are not only harming the canines physical and psychological state, but the workers and potential adoptive families physical and psychological state as well. The workers psychological state could affect the care provided to the canines. These loud noise exposures also have the potential to reduce the amount of time that potential adoptive families spend in the facility. This can result in less dogs being adopted and more time being exposed to excessive sound levels.

To reduce the level of noise exposure poses a little more difficulty because the majority of the noise is coming from the canines (barking), but structural changes can be made to the facilities in order to reduce the noise. Structural changes could include how many dogs are put in one area, more absorbing material rather than metal cages and cement walls and floors, and possibly in the future use of hearing protection devices (HPD) for the canines. All of these structural changes would also benefit the humans involved as well as the use of HPD's (ear plugs).

References

- Adoga, AA; Obindo, TJ (2013). The Association between tinnitus and mental illness, mental disorders - Theoretical and Empirical Perspectives. InTech. ISBN 978-953-51-0919-8.
- Kryter, Karl D. (1994). The handbook of hearing and the effects of noise: physiology, psychology, and public health. Boston: Academic Press. ISBN 0-12-427455-2.
- Halpern, David (1995). Mental health and the built environment: more than bricks and mortar?. Taylor & Francis. ISBN 0-7484-0235-7.
- Stellman, Jeanne Mager (1998). Encyclopedia of occupational health and safety. International Labour Organization. ISBN 92-2-109203-8.
- Williams, Luanne Kemp; Langley, Rick L. (2000). Environmental Health Secrets. Philadelphia: Elsevier Health Sciences. ISBN 1-56053-408-7.
- "Hearing Impairment Among Noise-Exposed Workers — United States, 2003–2012 | MMWR". www.cdc.gov. Retrieved 2016-04-21.
- "Ototoxic chemicals - chemicals that result in hearing loss". Department of Commerce Western Australia. Retrieved 2016-03-28.
- "Noise controls (Engineering, Administrative, PPE) | Mining Safety & Health Resource Center". miningsh.arizona.edu. Retrieved 2016-03-30.
- Kardous, C; Themann, C; Morata, T; Reynolds, J; Afunuh, S (2015). "Workplace Solutions: Reducing the Risk of Hearing Disorders among Musicians" (PDF). National Institute for Occupational Safety and Health. Retrieved July 12, 2016.
- "Occupationally-Induced Hearing Loss: Employers and Employees in Manufacturing need your healp" (PDF). National Institute for Occupational Safety and Health. March 2010. Retrieved July 12, 2016.
- "Worker Safety Series: Protecting Yourself from Noise in Construction" (PDF). Occupational Safety and Health Administration. Retrieved July 12, 2016.
- "Results and Discussion - Effects - Noise Effect On Wildlife - Noise - Environment - FHWA". www.fhwa.dot.gov. Retrieved 2015-12-21.

5

Noise Control: Tools and Technologies

There are few technologies that can prevent noise pollution. Some of them are construction of soundproof rooms for noisy machines, absorption and vibration isolation. Excessive noise can affect psychological health, cause sleeping disorders and can also cause death in animals. This chapter has been carefully written to provide an easy understanding of noise control.

Noise Control

Noise control or noise mitigation is a set of strategies to reduce noise pollution or to reduce the impact of that noise, whether outdoors or indoors.

Overview

The main areas of noise mitigation or abatement are: transportation noise control, architectural design, urban planning through zoning codes, and occupational noise control. Roadway noise and aircraft noise are the most pervasive sources of environmental noise. Social activities may generate noise levels that consistently affect the health of populations residing in or occupying areas, both indoor and outdoor, near entertainment venues that feature amplified sounds and music that present significant challenges for effective noise mitigation strategies.

Multiple techniques have been developed to address interior sound levels, many of which are encouraged by local building codes; in the best case of project designs, planners are encouraged to work with design engineers to examine trade-offs of roadway design and architectural design. These techniques include design of exterior walls, party walls, and floor and ceiling assemblies; moreover, there are a host of specialized means for damping reverberation from special-purpose rooms such as auditoria, concert halls, entertainment and social venues, dining areas, audio recording rooms, and meeting rooms.

Many of these techniques rely upon materials science applications of constructing sound baffles or using sound-absorbing liners for interior spaces. Industrial noise control is really a subset of interior architectural control of noise, with emphasis upon specific methods of sound isolation from industrial machinery and for protection of workers at their task stations.

Sound masking is the active addition of noise to reduce the annoyance of certain sounds; the opposite of soundproofing.

Basic Technologies

- Sound insulation: prevent the transmission of noise by the introduction of a mass barrier. Common materials have high-density properties such as brick, thick glass, concrete, metal etc.
- Sound absorption: a porous material which acts as a 'noise sponge' by converting the sound energy into heat within the material. Common sound absorption materials include decoupled lead-based tiles, open cell foams and fiberglass
- Vibration damping: applicable for large vibrating surfaces. The damping mechanism works by extracting the vibration energy from the thin sheet and dissipating it as heat. A common material is sound deadened steel.
- Vibration isolation: prevents transmission of vibration energy from a source to a receiver by introducing a flexible element or a physical break. Common vibration isolators are springs, rubber mounts, cork etc.

Roadways

This noise abatement wall in The Netherlands has a transparent section at the driver's eye-level to reduce the visual impact for road users.

Source control in roadway noise has provided little reduction in vehicle noise, except for the development of the hybrid vehicle; nevertheless, hybrid use will need to attain a market share of roughly fifty percent to have a major impact on noise source reduction of city streets. Highway noise is today less affected by motor type, since the effects in higher speed are aerodynamic and tire noise related. Other contributions to reduction of noise at the source are: improved tire tread designs for trucks in the 1970s, better shielding of diesel stacks in the 1980s, and local vehicle regulation of unmuffled vehicles.

The most fertile areas for roadway noise mitigation are in urban planning decisions, roadway design, noise barrier design, speed control, surface pavement selection and truck restrictions. Speed control is effective since the lowest sound emissions arise from vehicles moving smoothly at 30 to 60 kilometres per hour. Above that range, sound emissions double with each five miles per hour of speed. At the lowest speeds, braking and (engine) acceleration noise dominates.

Selection of road surface pavement can make a difference of a factor of two in sound levels, for the speed regime above 30 kilometres per hour. Quieter pavements are porous with a negative surface texture and use medium to small aggregates; the loudest pavements have a transversely grooved surface, and/or a positive surface texture and use larger aggregates. Surface friction and roadway safety are important considerations as well for pavement decisions.

When designing new urban freeways or arterials, there are numerous design decisions regarding alignment and roadway geometrics. Use of a computer model to calculate sound levels has become standard practice since the early 1970s. In this way exposure of sensitive receptors to elevated sound levels can be minimized. An analogous process exists for urban mass transit systems and other rail transportation decisions. Early examples of urban rail systems designed using this technology were: Boston MBTA line expansions (1970s), San Francisco BART system expansion (1981), Houston METRORail system (1982), and the MAX Light Rail system in Portland, Oregon (1983).

Noise barriers can be applicable for existing or planned surface transportation projects. They are probably the single most effective weapon in retrofitting an existing roadway, and commonly can reduce adjacent land use sound levels by up to ten decibels. A computer model is required to design the barrier since terrain, micrometeorology and other locale specific factors make the endeavor a very complex undertaking. For example, a roadway in cut or strong prevailing winds can produce a setting where atmospheric sound propagation is unfavorable to any noise barrier.

Aircraft

A British Airways Airbus A321, on landing approach to London Heathrow Airport, showing proximity to homes.

As in the case of roadway noise, little progress has been made in quelling aircraft noise at the source, other than elimination of loud engine designs from the 1960s and earlier. Because of its velocity and volume, jet turbine engine exhaust noise defies reduction by any simple means.

The most promising forms of aircraft noise abatement are through land planning, flight operations restrictions and residential soundproofing. Flight restrictions can take the form of preferred runway use, departure flight path and slope, and time-of-day restrictions. These tactics are sometimes controversial since they can impact aircraft safety, flying convenience and airline economics.

In 1979, the US Congress authorized the FAA to devise technology and programs to attempt to insulate homes near airports. While this obviously does not aid the exterior environment, the program has been effective for residential and school interiors. Some of the first airports at which the technology was applied were San Francisco International Airport, Seattle-Tacoma International Airport, John Wayne International Airport and San Jose International Airport in California.

The underlying technology is a computer model which simulates the propagation of aircraft noise and its penetration into buildings. Variations in aircraft types, flight patterns and local meteorology can be analyzed along with benefits of alternative building retrofit strategies such as roof upgrading, window glazing improvement, fireplace baffling, caulking construction seams and other measures. The computer model allows cost effectiveness evaluations of a host of alternative strategies.

In Canada, Transport Canada prepares noise exposure forecasts (NEF) for each airport, using a computer model similar to that used in US residential land development is discouraged within high impact areas identified by the forecast.

In 1998, the flight paths in all of Scandinavia were changed as the new Oslo-Gardermoen Airport was opened. These new paths were straighter, reducing fuel use, and disturbing fewer people, however, vociferous protests came from people near the new paths who had not been disturbed before, and they took legal action (NIMBY effect).

Architectural Solutions

Architectural acoustics noise control practices include: interior sound reverberation reduction, inter-room noise transfer mitigation and exterior building skin augmentation.

In the case of construction of new (or remodeled) apartments, condominiums, hospitals, and hotels, many states and cities have stringent building codes with requirements of acoustical analysis, in order to protect building occupants. With regard to exterior noise, the codes usually require measurement of the exterior acoustic environment in order to determine the performance standard required for exterior building skin design. The architect can work with the acoustical scientist to arrive at the best cost-effective means of creating a quiet interior (normally 45 dBA). The most important elements of design of the building skin are usually: glazing (glass thickness, double pane design etc.), perforated metal (used internally or externally), roof material, caulking standards, chimney baffles, exterior door design, mail slots, attic ventilation ports, and mounting of through-the-wall air conditioners.

Sound treatment panels contrast with red curtains in a church meeting hall

Soundproof doors in a broadcast center

Acoustic ceiling tiles

Regarding sound generated inside the building, there are two principal types of transmission. Firstly, airborne sound travels through walls or floor and ceiling assemblies and can emanate from either human activities in adjacent living spaces or from mechanical noise within the building systems. Human activities might include voice, noise from amplified sound systems, or animal noise. Mechanical systems are elevator systems, boilers, refrigeration or air conditioning systems, generators and trash compactors. Aerodynamic sources include fans, pneumatics, and combustion. Noise control for aerodynamic sources include quiet air nozzles, pneumatic silencers and quiet fan technology. Since many mechanical sounds are inherently loud, the principal design element is to require the wall or ceiling assembly to meet certain performance standards, (typically Sound transmission class of 50), which allows considerable attenuation of the sound level reaching occupants.

The second type of interior sound is called Impact Insulation Class (IIC) transmission. This effect arises not from airborne transmission, but rather from transmission of sound through the building itself. The most common perception of IIC noise is from footfall of occupants in living spaces above. Low frequency noise is transferred easily through the ground and buildings. This type of noise is more difficult to abate, but consideration must be given to isolating the floor assembly above or hanging the lower ceiling on resilient channel.

Both of the transmission effects noted above may emanate either from building occupants or from building mechanical systems such as elevators, plumbing systems or heating, ventilating and air conditioning units. In some cases it is merely necessary to specify the best available quieting technology in selecting such building hardware. In other cases shock mounting of systems to control vibration may be in order. In the case of plumbing systems there are specific protocols developed, especially for water supply lines, to create isolation clamping of pipes within building walls. In the case of central air systems, it is important to baffle any ducts that could transmit sound between different building areas.

Designing special-purpose rooms has more exotic challenges, since these rooms may have requirements for unusual features such as concert performance, sound studio recording, lecture halls. In these cases reverberation and reflection must be analyzed in order to not only quiet the rooms, but to prevent echo effects from occurring. In these situations special sound baffles and sound absorptive lining materials may be specified to dampen unwanted effects.

Materials

Acoustical wall and ceiling panels can be constructed of many different materials and finishes. The ideal acoustical panels are those without a face or finish material that interferes with the acoustical infill or substrate. Fabric covered panels are one way to maximize the acoustical absorption. The finish material is used to cover over the acoustical substrate. Mineral fiber board, or Micore, is a commonly used acoustical substrate. Finish materials often consist of fabric, wood or metal. Fabric can be wrapped around substrates to create what is referred to as a "pre-fabricated panel" if laid onto a wall, and require no modifications. Such fabrics are generally acoustically 'transparent, meaning that they do not impede a sound wave.

Prefabricated panels are limited to the size of the subas "on-site acoustical wall panels" This is constructed by "framing" the perimeter track into shape, infilling the acoustical substrate and then stretching and tucking the fabric into the perimeter frame system. On-site wall panels can be constructed to work around door frames, baseboard, or any other intrusion. Large panels (generally greater than 50 feet) can be created on walls and ceilings with this method.

Double-glazed and thicker windows can also prevent sound transmission from the outdoors.

Industrial

This situation classically is thought to involve primarily manufacturing settings where industrial machinery produces intense sound levels, not uncommonly in the 75 to 85 decibel range. While this circumstance is the most dramatic, there are many other office type environments where sound levels may lie in the range of 70 to 75 decibels, entirely composed of office equipment, music, public address systems, and even exterior noise intrusion. The latter environments can also produce noise health effects provided that exposures are long term.

In the case of industrial equipment, the most common techniques for noise protection of workers consist of shock mounting source equipment, creation of acrylic glass or other solid barriers, and provision of ear protection equipment. In certain cases the machinery itself can be re-designed to operate in a manner less prone to produce grating, grinding, frictional, or other motions that induce sound emissions. In recent years, Buy Quiet programs and initiatives have arisen in an effort to combat occupational noise exposures. These programs promote the purchase of quieter tools and equipment and encourage manufacturers to design quieter equipment.

In the case of more conventional office environments, the techniques in architectural acoustics discussed above may apply. Other solutions may involve researching the quietest models of office equipment, particularly printers and photocopy machines. Impact printers and other equipment were often fitted with "acoustic hoods", enclosures to reduce emitted noise. One source of annoying, if not loud, sound level emissions are lighting fixtures (notably older fluorescent globes). These fixtures can be retrofitted or analyzed to see whether over-illumination is present, a common office environment issue. If over-illumination is occurring, de-lamping or reduced light bank usage may apply.

Commercial

Reductions in cost of technology have allowed noise control technology to be used not only in performance facilities and recording studios, but also in noise-sensitive small businesses such as restaurants. Acoustically absorbent materials such as fiberglass duct liner, wood fiber panels and recycled denim jeans serve as artwork-bearing canvasses in environments in which aesthetics are important.

Using a combination of sound absorption materials, arrays of microphones and speakers, and a digital processor, a restaurant operator can use a tablet computer to selectively control noise levels at different places in the restaurant: the microphone arrays pick up sound and send it to the digital processor, which controls the speakers to output sound signals on command.

Urban Planning

Communities may use zoning codes to isolate noisy urban activities from areas that should be protected from such unhealthy exposures and to establish noise standards in areas that may not be conducive to such isolation strategies. Because low-income neighborhoods are often at greater risk of noise pollution, the establishment of such zoning codes is often an environmental justice issue. Mixed use areas present especially difficult conflicts that require special attention to the need to protect people from the harmful effects of noise pollution. Noise is generally one consideration in an environmental impact statement, if applicable (such as transportation system construction).

Soundproofing

An anechoic chamber, showing acoustic damping tiles used for sound absorption.

Sound reflection board

Soundproofing is any means of reducing the sound pressure with respect to a specified sound source and receptor. There are several basic approaches to reducing sound: increasing the distance between source and receiver, using noise barriers to reflect or absorb the energy of the sound waves, using damping structures such as sound baffles, or using active antinoise sound generators.

Two distinct soundproofing problems may need to be considered when designing acoustic treatments - to improve the sound within a room, and reduce sound leakage to/from adjacent rooms or outdoors. Acoustic quieting, noise mitigation, and noise control can be used to limit unwanted noise. Soundproofing can suppress unwanted indirect sound waves such as reflections that cause echoes and resonances that cause reverberation. Soundproofing can reduce the transmission of unwanted direct sound waves from the source to an involuntary listener through the use of distance and intervening objects in the sound path.

Distance

The energy density of sound waves decreases as they spread out, so that increasing the distance between the receiver and source results in a progressively lesser intensity of sound at the receiver. In a normal three-dimensional setting, with a point source and point receptor, the intensity of sound waves will be attenuated according to the inverse square of the distance from the source.

Damping

Damping means to reduce resonance in the room, by absorption or redirection (reflection or diffusion). Absorption will reduce the overall sound level, whereas redirection makes unwanted sound harmless or even beneficial by reducing coherence. Damping can reduce the acoustic resonance in the air, or mechanical resonance in the structure of the room itself or things in the room. When constructing a vehicle which includes soundproofing, a panel dampening material is fitted which reduces the vibration of the vehicles body panels when they are excited by one of the many high energy sound sources caused when the vehicle is in use.

Absorption

Absorbing sound spontaneously converts part of the sound energy to a very small amount of heat in the intervening object (the absorbing material), rather than sound being transmitted or reflected. There are several ways in which a material can absorb sound. The choice of sound absorbing material will be determined by the frequency distribution of noise to be absorbed and the acoustic absorption profile required.

Porous Absorbers

Porous absorbers, typically open cell rubber foams or melamine sponges, absorb noise by friction within the cell structure.

Porous open cell foams are highly effective noise absorbers across a broad range of medium-high frequencies. Performance is less impressive at low frequencies.

The exact absorption profile of a porous open cell foam will be determined by a number of factors including the following:

- Cell size
- Tortuosity
- Porosity
- Material thickness
- Material density

Resonant Absorbers

Resonant panels, Helmholtz resonators and other resonant absorbers work by damping a sound wave as they reflect it.

Unlike porous absorbers, resonant absorbers are most effective at low-medium frequencies and the absorption of resonant absorbers is always matched to a narrow frequency range.

Reflection

In an outdoor environment such as highway engineering, embankments or panelling are often used to reflect sound upwards into the sky.

Diffusion

If a specular reflection from a hard flat surface is giving a problematic echo then an acoustic diffuser may be applied to the surface. It will scatter sound in all directions.

Room Within a Room

A room within a room (RWAR) is one method of isolating sound and preventing it from transmitting to the outside world where it may be undesirable.

Most vibration / sound transfer from a room to the outside occurs through mechanical means. The vibration passes directly through the brick, woodwork and other solid structural elements. When it meets with an element such as a wall, ceiling, floor or window, which acts as a sounding board, the vibration is amplified and heard in the second space. A mechanical transmission is much faster, more efficient and may be more readily amplified than an airborne transmission of the same initial strength.

The use of acoustic foam and other absorbent means is less effective against this transmitted vibration. The user is advised to break the connection between the room that contains the noise source and the outside world. This is called acoustic decoupling. Ideal decoupling involves eliminating vibration transfer in both solid materials and in the air, so air-flow into the room is often controlled. This has safety implications: inside decoupled space, proper ventilation must be assured, and gas heaters cannot be used.

Noise Cancellation

Noise cancellation generators for active noise control are a relatively modern innovation. A microphone is used to pick up the sound that is then analyzed by a computer; then, sound waves with opposite polarity (180° phase at all frequencies) are output through a speaker, causing destructive interference and cancelling much of the noise.

Residential Soundproofing

Residential soundproofing aims to decrease or eliminate the effects of exterior noise. The main focus of residential soundproofing in existing structures is the windows. Curtains can be used to damp sound either through use of heavy materials or through the use of air chambers known as honeycombs. Single-, double- and triple-honeycomb designs achieve relatively greater degrees of sound damping. The primary soundproofing limit of curtains is the lack of a seal at the edge of the curtain, although this may be alleviated with the use of sealing features, such as hook and loop fastener, adhesive, magnets, or other materials. Double-pane windows achieve somewhat greater sound damping than single-pane windows. Significant noise reduction can be achieved by installing a second interior window. In this case the exterior window remains in place while a slider or hung window is installed within the same wall openings.

Commercial Soundproofing

Acoustical sound panels on the ceiling in a restaurant

Restaurants, schools, office businesses, and health care facilities use architectural acoustics to reduce noise for their customers.

Commercial businesses sometimes use soundproofing technology, especially when they are an open office design. There are many reasons why a business might implement soundproofing for their office. One of the biggest hindrances in worker productivity are the distracting noises that comes from people talking such as on the phone, or with their co-workers and boss. Noise soundproofing is important in mitigating people from losing their concentration and focus from their work project. It is also important to keep confidential conversations secure to the intended listeners.

When trying to find places to install soundproofing, acoustic panels should be installed in office areas where lots of traffic corridor or circulation pathways, and open work areas that are connected.

In the US, OSHA has requirements regulating the length of exposure of workers to certain levels of noise.

Successful acoustic panel installations rely on three strategies and techniques: 1) absorb sound, 2) block sound transmission from one place to another, and 3) cover through masking of the sound.

Automotive Soundproofing

Automotive soundproofing aims to decrease or eliminate the effects of exterior noise, primarily engine, exhaust and tire noise across a wide frequency range. The automotive environment limits the thickness of materials that can be used, but combinations of dampers, barriers, and absorbers are common. Common materials include felt, foam, polyester, and Polypropylene blend materials. Waterproofing may be necessary based on materials used.

There are many complex noises created within vehicles which change with the driving environment and speed at which the vehicle travels. Significant noise reductions of up to 8 dB can be achieved by installing a combination of different types of materials.

Noise barriers as Exterior Soundproofing

Since the early 1970s, it has become common practice in the United States and other industrialized countries to engineer noise barriers along major highways to protect adjacent residents from intruding roadway noise. Engineering techniques have been developed to predict an effective geometry for the noise barrier design in a particular real world situation. Noise barriers may be constructed of wood, masonry, earth or a combination thereof. One of the earliest noise barrier designs was in Arlington, Virginia adjacent to Interstate 66, stemming from interests expressed by the Arlington

Coalition on Transportation. Possibly the earliest scientifically designed and published noise barrier construction was in Los Altos, California in 1970.

Absorption (Acoustics)

Acoustic absorption refers to the process by which a material, structure, or object takes in sound energy when sound waves are encountered, as opposed to reflecting the energy. Part of the absorbed energy is transformed into heat and part is transmitted through the absorbing body. The energy transformed into heat is said to have been 'lost'. When sound from a loudspeaker collides with the walls of a room part of the sound's energy is reflected, part is transmitted, and part is absorbed into the walls. As the waves travel through the wall they deform the material thereof (just like they deformed the air before). This deformation causes mechanical losses via conversion of part of the sound energy into heat, resulting in acoustic attenuation, mostly due to the wall's viscosity. Similar attenuation mechanisms apply for the air and any other medium through which sound travels. The fraction of sound absorbed is governed by the acoustic impedances of both media and is a function of frequency and the incident angle. Size and shape can influence the sound wave's behavior if they interact with its wavelength, giving rise to wave phenomena such as standing waves and diffraction.

Acoustic absorption is of particular interest in soundproofing. Soundproofing aims to absorb as much sound energy (often in particular frequencies) as possible converting it into heat or transmitting it away from a certain location.

In general, soft, pliable, or porous materials (like cloths) serve as good acoustic insulators - absorbing most sound, whereas dense, hard, impenetrable materials (such as metals) reflect most.

How well a room absorbs sound is quantified by the effective absorption area of the walls, also named total absorption area. This is calculated using its dimensions and the absorption coefficients of the walls. The total absorption is expressed in Sabins and is useful in, for instance, determining the reverberation time of auditoria. Absorption coefficients can be measured using a reverberation room, which is the opposite of an anechoic chamber.

Applications

Acoustic absorption is critical in areas such as:

- Soundproofing
- Sound recording and reproduction
- Loudspeaker design

- Acoustic transmission lines
- Room acoustics
- Architectural acoustics
- Sonar
- Noise Barrier Walls

Anechoic Chamber

An acoustic anechoic chamber is a room designed to absorb as much sound as possible. The walls consist of a number of baffles with highly absorptive material arranged in such a way that the fraction of sound they do reflect is directed towards another baffle instead of back into the room. This makes the chamber almost devoid of echos which is useful for measuring the sound pressure level of a source and for various other experiments and measurements.

Anechoic chambers are expensive for several reasons and are therefore not common.

They must be isolated from outside influences (eg, planes, trains, automobiles, snowmobiles, elevators, pumps, ...; indeed any source of sound which may interfere with measurements inside the chamber) and they must be physically large. The first, environmental isolation, requires in most cases specially constructed, nearly always massive, and likewise thick, walls, floors, and ceilings. Such chambers are often built as spring supported isolated rooms within a larger building. The National Research Council in Canada has a modern anechoic chamber, and has posted a video on the Web, noting these as well as other constructional details. Doors must be specially made, sealing for them must be acoustically complete (no leaks around the edges), ventilation (if any) carefully managed, and lighting chosen to be silent.

The second requirement follows in part from the first and from the necessity of preventing reverberation inside the room from, say, a sound source being tested. Preventing echoes is almost always done with absorptive foam wedges on walls, floors and ceilings, and if they are to be effective at low frequencies, these must be physically large; the lower the frequencies to be absorbed, the larger they must be.

An anechoic chamber must therefore be large to accommodate those absorbers and isolation schemes, but still allow for space for experimental apparatus and units under test.

Electrical and Mechanical Analogy

The energy dissipated within a medium as sound travels through it is analogous to the energy dissipated in electrical resistors or that dissipated in mechanical dampers for mechanical motion transmission systems. All three are equivalent to the resistive part of a system of resistive and reactive elements. The resistive elements dissipate energy (irreversibly into heat) and the reactive elements store and release energy (reversibly, neglecting small losses). The reactive parts of an acoustic medium are determined by its bulk modulus and its density, analogous to respectively an electrical capacitor and an electrical inductor, and analogous to, respectively, a mechanical spring attached to a mass.

Note that since dissipation solely relies on the resistive element it is independent of frequency. In practice however the resistive element varies with frequency. For instance, vibrations of most materials change their physical structure and so their physical properties; the result is a change in the 'resistance' equivalence. Additionally, the cycle of compression and rarefaction exhibits hysteresis of pressure waves in most materials which is a function of frequency, so for every compression there is a rarefaction, and the total amount of energy dissipated due to hysteresis changes with frequency. Furthermore some materials behave in a non-Newtonian way, which causes their viscosity to change with the rate of shear strain experienced during compression and rarefaction; again, this varies with frequency. Gasses and liquids generally exhibit less hysteresis than solid materials (eg, sound waves cause adiabatic compression and rarefaction) and behave in a, mostly, Newtonian way.

Combined, the resistive and reactive properties of an acoustic medium form the acoustic impedance. The behaviour of sound waves encountering a different medium is dictated by the differing acoustic impedances. As with electrical impedances, there are matches and mismatches and energy will be transferred for certain frequencies (up to nearly 100%) whereas for others it could be mostly reflected (again, up to very large percentages).

In amplifier and loudspeaker design electrical impedances, mechanical impedances, and acoustic impedances of the system have to be balanced such that the frequency and phase response least alter the reproduced sound across a very broad spectrum whilst still producing adequate sound levels for the listener. Modelling acoustical systems using the same (or similar) techniques long used in electrical circuits gave acoustical designers a new and powerful design tool.

Vibration Isolation

Vibration isolation is the process of isolating an object, such as a piece of equipment, from the source of vibrations.

Vibration is undesirable in many domains, primarily engineered systems and habitable spaces, and methods have been developed to prevent the transfer of vibration to such systems. Vibrations propagate via mechanical waves and certain mechanical linkages conduct vibrations more efficiently than others. Passive vibration isolation makes use of materials and mechanical linkages that absorb and damp these mechanical waves. Active vibration isolation involves sensors and actuators that produce destructive interference that cancels-out incoming vibration.

Passive Isolation

"Passive vibration isolation" refers to vibration isolation or mitigation of vibrations by passive techniques such as rubber pads or mechanical springs, as opposed to "active vibration isolation" or "electronic force cancellation" employing electric power, sensors, actuators, and control systems.

Passive vibration isolation is a vast subject, since there are many types of passive vibration isolators used for many different applications. A few of these applications are for industrial equipment such as pumps, motors, HVAC systems, or washing machines; isolation of civil engineering structures from earthquakes (base isolation), sensitive laboratory equipment, valuable statuary, and high-end audio.

A basic understanding of how passive isolation works, the more common types of passive isolators, and the main factors that influence the selection of passive isolators:

Common Passive Isolation Systems

Pneumatic or air isolators

> These are bladders or canisters of compressed air. A source of compressed air is required to maintain them. Air springs are rubber bladders which provide damping as well as isolation and are used in large trucks. Some pneumatic isolators can attain low resonant frequencies and are used for isolating large industrial equipment. Air tables consist of a working surface or optical surface mounted on air legs. These tables provide enough isolation for laboratory instrument under some conditions. Air systems may leak under vacuum conditions. The air container can interfere with isolation of low-amplitude vibration.

Mechanical springs and spring-dampers

> These are heavy-duty isolators used for building systems and industry. Sometimes they serve as mounts for a concrete block, which provides further isolation.

Pads or sheets of flexible materials such as elastomers, rubber, cork, dense foam and laminate materials.

> Elastomer pads, dense closed cell foams and laminate materials are often used under heavy machinery, under common household items, in vehicles and even under higher performing audio systems.

Molded and bonded rubber and elastomeric isolators and mounts

These are often used as machinery mounts or in vehicles. They absorb shock and attenuate some vibration.

Negative-stiffness isolators

Negative-stiffness isolators are less common than other types and have generally been developed for high-level research applications such as gravity wave detection. Lee, Goverdovskiy, and Temnikov (2007) proposed a negative-stiffness system for isolating vehicle seats.

The focus on negative-stiffness isolators has been on developing systems with very low resonant frequencies (below 1 Hz), so that low frequencies can be adequately isolated, which is critical for sensitive instrumentation. All higher frequencies are also isolated. Negative-stiffness systems can be made with low stiction, so that they are effective in isolating low-amplitude vibrations.

Negative-stiffness mechanisms are purely mechanical and typically involve the configuration and loading of components such as beams or inverted pendulums. Greater loading of the negative-stiffness mechanism, within the range of its operability, decreases the natural frequency.

Wire rope isolators

These isolators are durable and can withstand extreme environments. They are often used in military applications.

Base isolators for seismic isolation of buildings, bridges, etc.

Base isolators made of layers of neoprene and steel with a low horizontal stiffness are used to lower the natural frequency of the building. Some other base isolators are designed to slide, preventing the transfer of energy from the ground to the building.

Tuned mass dampers

Tuned mass dampers reduce the effects of harmonic vibration in buildings or other structures. A relatively small mass is attached in such a way that it can dampen out a very narrow band of vibration of the structure.

Do it Yourself Isolators

In less sophisticated solutions, bungee cords can be used as a cheap isolation system which may be effective enough for some applications. The item to be isolated is suspended from the bungee cords. This is difficult to implement without a danger of the isolated item falling. Tennis balls cut in half have been used under washing machines and other items with some success.

How Passive Isolation Works

A passive isolation system, such as a shock mount, in general contains mass, spring, and damping elements and moves as a harmonic oscillator. The mass and spring stiffness dictate a natural frequency of the system. Damping causes energy dissipation and has a secondary effect on natural frequency.

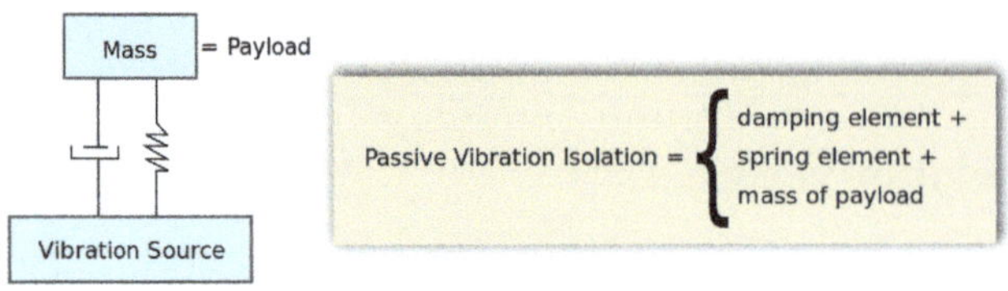

Passive Vibration Isolation

Every object on a flexible support has a fundamental natural frequency. When vibration is applied, energy is transferred most efficiently at the natural frequency, somewhat efficiently below the natural frequency, and with increasing inefficiency (decreasing efficiency) above the natural frequency. This can be seen in the transmissibility curve, which is a plot of transmissibility vs. frequency.

Here is an example of a transmissibility curve. Transmissibility is the ratio of vibration of the isolated surface to that of the source. Vibrations are never completely eliminated, but they can be greatly reduced. The curve below shows the typical performance of a passive, negative-stiffness isolation system with a natural frequency of 0.5 Hz. The general shape of the curve is typical for passive systems. Below the natural frequency, transmissibility hovers near 1. A value of 1 means that vibration is going through the system without being amplified or reduced. At the resonant frequency, energy is transmitted efficiently, and the incoming vibration is amplified. Damping in the system limits the level of amplification. Above the resonant frequency, little energy can be transmitted, and the curve rolls off to a low value. A passive isolator can be seen as a mechanical low-pass filter for vibrations.

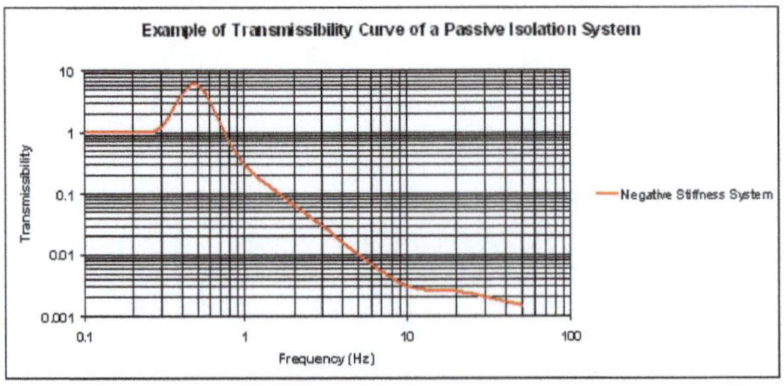

negative-stiffness transmissibility

In general, for any given frequency above the natural frequency, an isolator with a lower natural frequency will show greater isolation than one with a higher natural frequency. The best isolation system for a given situation depends on the frequency, direction, and magnitude of vibrations present and the desired level of attenuation of those frequencies.

All mechanical systems in the real world contain some amount of damping. Damping dissipates energy in the system, which reduces the vibration level which is transmitted at the natural frequency. The fluid in automotive shock absorbers is a kind of damper, as is the inherent damping in elastomeric (rubber) engine mounts.

Damping is used in passive isolators to reduce the amount of amplification at the natural frequency. However, increasing damping tends to reduce isolation at the higher frequencies. As damping is increased, transmissibility roll-off decreases. This can be seen in the chart below.

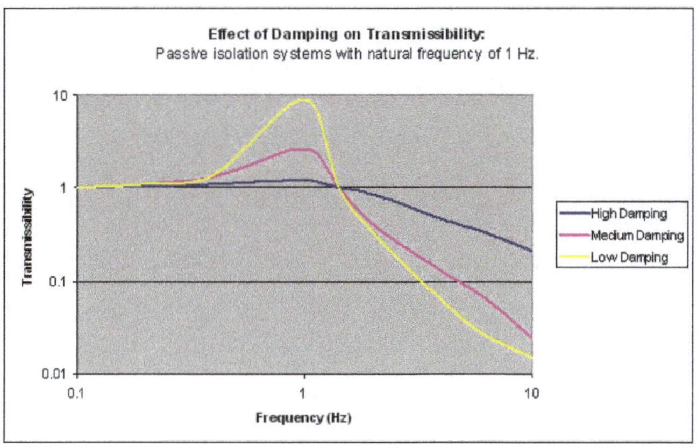

Damping effect on transmissibility

Passive isolation operates in both directions, isolating the payload from vibrations originating in the support, and also isolating the support from vibrations originating in the payload. Large machines such as washers, pumps, and generators, which would cause vibrations in the building or room, are often isolated from the floor. However, there are a multitude of sources of vibration in buildings, and it is often not possible to isolate each source. In many cases, it is most efficient to isolate each sensitive instrument from the floor. Sometimes it is necessary to implement both approaches.

Factors Influencing the Selection of Passive Vibration Isolators

1. Characteristics of item to be isolated
    - Size: The dimensions of the item to be isolated help determine the type of isolation which is available and appropriate. Small objects may use only one isolator, while larger items might use a multiple-isolator system.
    - Weight: The weight of the object to be isolated is an important factor in choosing the correct passive isolation product. Individual passive isolators are designed to be used with a specific range of loading.

    - Movement: Machines or instruments with moving parts may affect isolation systems. It is important to know the mass, speed, and distance traveled of the moving parts.

2. Operating Environment

    - Industrial: This generally entails strong vibrations over a wide band of frequencies and some amount of dust.
    - Laboratory: Labs are sometimes troubled by specific building vibrations from adjacent machinery, foot traffic, or HVAC airflow.
    - Indoor or outdoor: Isolators are generally designed for one environment or the other.
    - Corrosive/non-corrosive: Some indoor environments may present a corrosive danger to isolator components due to the presence of corrosive chemicals. Outdoors, water and salt environments need to be considered.
    - Clean room: Some isolators can be made appropriate for clean room.
    - Temperature: In general, isolators are designed to be used in the range of temperatures normal for human environments. If a larger range of temperatures is required, the isolator design may need to be modified.
    - Vacuum: Some isolators can be used in a vacuum environment. Air isolators may have leakage problems. Vacuum requirements typically include some level of clean room requirement and may also have a large temperature range.
    - Magnetism: Some experimentation which requires vibration isolation also requires a low-magnetism environment. Some isolators can be designed with low-magnetism components.
    - Acoustic noise: Some instruments are sensitive to acoustic vibration. In addition, some isolation systems can be excited by acoustic noise. It may be necessary to use an acoustic shield. Air compressors can create problematic acoustic noise, heat, and airflow.
    - Static or dynamic loads: This distinction is quite important as isolators are designed for a certain type and level of loading.

Static loading

is basically the weight of the isolated object with low-amplitude vibration input. This is the environment of apparently stationary objects such as buildings (under normal conditions) or laboratory instruments.

Dynamic loading

involves accelerations and larger amplitude shock and vibration. This environment is present in vehicles, heavy machinery, and structures with significant movement.

3. Cost:

   - Cost of providing isolation: Costs include the isolation system itself, whether it is a standard or custom product; a compressed air source if required; shipping from manufacturer to destination; installation; maintenance; and an initial vibration site survey to determine the need for isolation.

   - Relative costs of different isolation systems: Inexpensive shock mounts may need to be replaced due to dynamic loading cycles. A higher level of isolation which is effective at lower vibration frequencies and magnitudes generally costs more. Prices can range from a few dollars for bungee cords to millions of dollars for some space applications.

4. Adjustment: Some isolation systems require manual adjustment to compensate for changes in weight load, weight distribution, temperature, and air pressure, whereas other systems are designed to automatically compensate for some or all of these factors.

5. Maintenance: Some isolation systems are quite durable and require little or no maintenance. Others may require periodic replacement due to mechanical fatigue of parts or aging of materials.

6. Size Constraints: The isolation system may have to fit in a restricted space in a laboratory or vacuum chamber, or within a machine housing.

7. Nature of vibrations to be isolated or mitigated

   - Frequencies: If possible, it is important to know the frequencies of ambient vibrations. This can be determined with a site survey or accelerometer data processed through FFT analysis.

   - Amplitudes: The amplitudes of the vibration frequencies present can be compared with required levels to determine whether isolation is needed. In addition, isolators are designed for ranges of vibration amplitudes. Some isolators are not effective for very small amplitudes.

   - Direction: Knowing whether vibrations are horizontal or vertical can help to target isolation where it is needed and save money.

8. Vibration specifications of item to be isolated: Many instruments or machines have manufacturer-specified levels of vibration for the operating environment. The manufacturer may not guarantee the proper operation of the instrument if vibration exceeds the spec.

Comparison of Passive Isolators

Type of Passive Isolation	Applications	Typical Natural Frequency
Air Isolators	Large industrial equipment, some optics and instruments	1.5 - 3 Hz, large systems customized to 0.5 Hz
Springs or spring dampers	Heavy loads, pumps, compressors	3 - 9 Hz
Elastomer or cork pads	Large high-load applications where isolation of medium to high frequency noise and vibration is required	3 - 40 Hz, depending on size of pad and load
Molded or bonded elastomer mounts	Machinery, instruments, vehicles, aviation	10 - 20+ Hz
Negative-stiffness isolators	Electron microscopes, sensitive instruments, optics and laser systems, cryogenic systems	0.17 - 2.5 Hz
Wire rope isolators	Machinery, instruments, vehicles, aviation	10 - 40+ Hz
Bungee cord isolators	Laboratory, home, etc.	Depends on type of cord and the mass they support
Base isolators	Buildings and large structures	Low, seismic frequencies
Tuned Mass Dampers	Buildings, large structures, aerospace	Any, but usually used at low frequencies

Negative-stiffness Vibration Isolator

Negative-Stiffness-Mechanism (NSM) vibration isolation systems offer a unique passive approach for achieving low vibration environments and isolation against sub-Hertz vibrations. "Snap-through" or "over-center" NSM devices are used to reduce the stiffness of elastic suspensions and create compact six-degree-of-freedom systems with low natural frequencies. Practical systems with vertical and horizontal natural frequencies as low as 0.2 to 0.5 Hz are possible. Electro-mechanical auto-adjust mechanisms compensate for varying weight loads and provide automatic leveling in multiple-isolator systems, similar to the function of leveling valves in pneumatic systems. All-metal systems can be configured which are compatible with high vacuums and other adverse environments such as high temperatures.

These isolation systems enable vibration-sensitive instruments such as scanning probe microscopes, micro-hardness testers and scanning electron microscopes to operate in severe vibration environments sometimes encountered, for example, on upper floors of buildings and in clean rooms. Such operation would not be practical with pneumatic isolation systems. Similarly, they enable vibration-sensitive instruments to produce better images and data than those achievable with pneumatic isolators.

The theory of operation of NSM vibration isolation systems is summarized, some typical systems and applications are described, and data on measured performance is presented. The theory of NSM isolation systems is explained in References 1 and 2. It is summarized briefly for convenience.

Vertical-motion Isolation

A vertical-motion isolator is shown . It uses a conventional spring connected to an NSM consisting of two bars hinged at the center, supported at their outer ends on pivots, and loaded in compression by forces P. The spring is compressed by weight W to the operating position of the isolator, as shown in Figure 1. The stiffness of the isolator is K=KS-KN where KS is the spring stiffness and KN is the magnitude of a negative-stiffness which is a function of the length of the bars and the load P. The isolator stiffness can be made to approach zero while the spring supports the weight W.

Horizontal-motion Isolation

A horizontal-motion isolator consisting of two beam-columns is illustrated in Figure. 2. Each beam-column behaves like two fixed-free beam columns loaded axially by a weight load W. Without the weight load the beam-columns have horizontal stiffness KS With the weight load the lateral bending stiffness is reduced by the "beam-column" effect. This behavior is equivalent to a horizontal spring combined with an NSM so that the horizontal stiffness is K=KS-KN, and KN is the magnitude of the beam-column effect. Horizontal stiffness can be made to approach zero by loading the beam-columns to approach their critical buckling load.

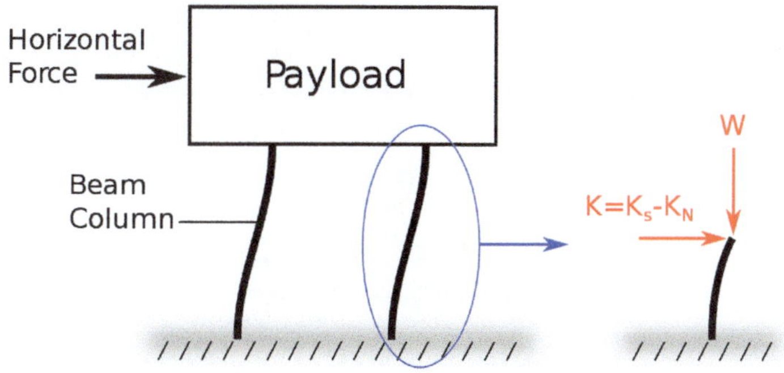

Six-degree-of-freedom (Six-DOF) Isolation

A six-DOF NSM isolator typically uses three isolators stacked in series: a tilt-motion isolator on top of a horizontal-motion isolator on top of a vertical-motion isolator. Figure 3 shows a schematic of a vibration isolation system consisting of a weighted platform supported by a single six-DOF isolator incorporating the isolators of Figures 1 and 2. Flexures are used in place of the hinged bars shown in Figure 1. A tilt flexure serves as the tilt-motion isolator. A vertical-stiffness adjustment screw is used to adjust the compression force on the negative-stiffness flexures thereby changing the vertical stiffness. A vertical load adjustment screw is used to adjust for varying weight loads by raising or lowering the base of the support spring to keep the flexures in their straight, unbent operating positions.

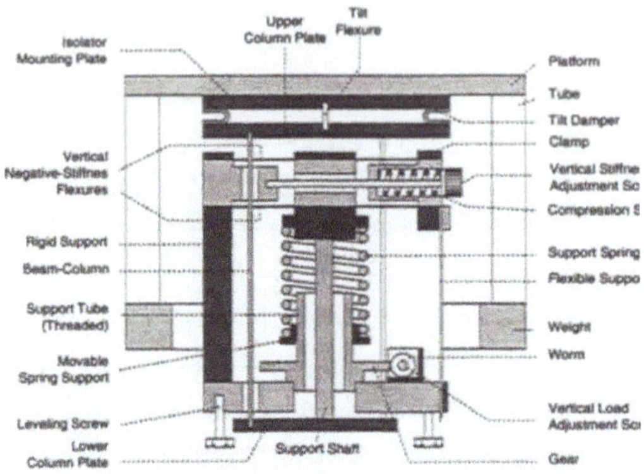

Figure 3 Schematic of Six-DOF Single-Isolator System

Vibration Isolation of Supporting Joint

The equipment or other mechanical components are necessarily linked to surrounding objects (the supporting joint - with the support; the unsupporting joint - the pipe duct or cable), thus presenting the opportunity for unwanted transmission of vibrations. Using a suitably designed vibration-isolator (absorber), vibration isolation of the supporting joint is realized. The accompanying illustration shows the attenuation of vibration levels, as measured before installation of the functioning gear on a vibration isolator as well as after installation, for a wide range of frequencies.

The Vibration Isolator

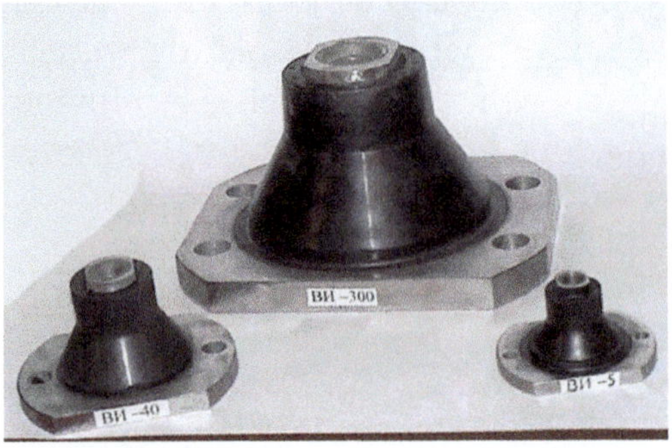

This is defined as a device that reflects and absorbs waves of oscillatory energy, extending from a piece of working machinery or electrical equipment, and with the desired effect being vibration insulation. The goal is to establish vibration isolation between a body transferring mechanical fluctuations and a supporting body (for example, between the machine and the foundation). The illustration shows a vibration isolator from the series «ВИ» (~"VI" in Roman characters), as used in shipbuilding in Russia, for example the submarine "St.Petersburg" (Lada). The depicted «ВИ» devices allow loadings ranging from 5, 40 and 300 kg. They differ in their physical sizes, but all share the same fundamental design. The structure consists of a rubber envelope that is internally reinforced by a spring. During manufacture, the rubber and the spring are intimately and permanently connected as a result of the vulcanization process that is integral to the processing of the crude rubber material. Under action of weight loading of the machine, the rubber envelope deforms, and the spring is compressed or stretched. Therefore, in the direction of the spring's cross section, twisting of the enveloping rubber occurs. The resulting elastic deformation of the rubber envelope results in very effective absorption of the vibration. This absorption is crucial to reliable vibration insulation, because it averts the potential for resonance effects. The amount of elastic deformation of the rubber largely dictates the magnitude of vibration absorption that can be attained; the entire device (including the spring itself) must be designed with this in mind. The design of the vibration isolator must also take into account potential exposure to shock loadings, in addition to the routine everyday vibrations. Lastly, the vibration isolator must also be designed for long-term durability as well as convenient integration into the environment in which it is to be used. Sleeves and flanges are typically employed in order to enable the vibration isolator to be securely fastened to the equipment and the supporting foundation.

Vibration Isolation of Unsupporting Joint

Vibration isolation of unsupporting joint is realized in the device named branch pipe a of isolating vibration.

Branch Pipe a of Isolating Vibration

Branch pipe a of isolating vibration is a part of a tube with elastic walls for reflection and absorption of waves of the oscillatory energy extending from the working pump over wall of the pipe duct. Is established between the pump and the pipe duct. On an illustration is presented the image a vibration-isolating branch pipe of a series «ВИПБ». In a structure is used the rubber envelope, which is reinforced by a spring. Properties of an envelope are similar envelope to an isolator vibration. Has the device reducing axial effort from action of internal pressure up to zero.

Subframe Isolation

Another technique used to increase isolation is to use an isolated subframe. This splits the system with an additional mass/spring/damper system. This doubles the high frequency attenuation rolloff, at the cost of introducing additional low frequency modes which may cause the low frequency behaviour to deteriorate. This is commonly used in the rear suspensions of cars with Independent Rear Suspension (IRS), and in the front subframes of some cars. The graph shows the force into the body for a subframe that is rigidly bolted to the body compared with the red curve that shows a compliantly mounted subframe. Above 42 Hz the compliantly mounted subframe is superior, but below that frequency the bolted in subframe is better.

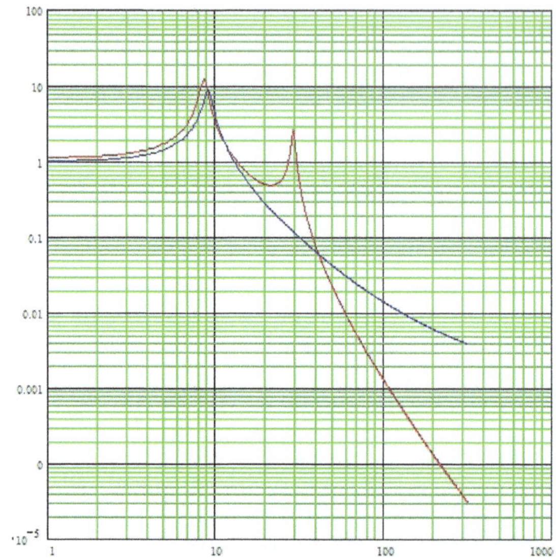

Subframe vibration isolation graph: force transmission on suspended body vs. frequency for rigidly and compliantly mounted subframes.

Active Isolation

Active vibration isolation systems contain, along with the spring, a feedback circuit which consists of a sensor (for example a piezoelectric accelerometer or a geophone), a controller, and an actuator. The acceleration (vibration) signal is processed by a control circuit and amplifier. Then it feeds the electromagnetic actuator, which amplifies the signal. As a result of such a feedback system, a considerably stronger suppression of vibrations is achieved compared to ordinary damping. Active isolation today is used for applications where structures smaller than a micrometer have to be produced or measured. A couple of companies produce active isolation products as OEM for research, metrology, lithography and medical systems. Another important application is the semiconductor industry. In the microchip production, the smallest structures today are below 20 nm, so the machines which produce and check them have to oscillate much less.

Sensors for Active Isolation

- Piezoelectric accelerometers and force sensors
- MEM accelerometers
- Geophones

- Proximity sensors
- Interferometers

Actuators for Active Isolation

- Linear motors
- Pneumatic actuators
- Piezoelectric motors

Acoustic Quieting

Acoustic quieting is the process of making machinery quieter by damping vibrations to prevent them from reaching the observer. Machinery vibrates, causing sound waves in air, hydroacoustic waves in water, and mechanical stresses in solid matter. Quieting is achieved by absorbing the vibrational energy or minimizing the source of the vibration. It may also be redirected away from the observer.

One of the major reasons for the development of acoustic quieting techniques was for making submarines difficult to detect by sonar. This military goal of the mid- and late-twentieth century allowed the technology to be adapted to many industries and products, such as computers (e.g. hard drive technology), automobiles (e.g. motor mounts), and even sporting goods (e.g. golf clubs).

Aspects of Acoustic Quieting

When the goal is acoustic quietening, a number of different aspects might be considered. Each aspect of acoustics can be taken alone or in concert so that the end result is that the reception of noise by the observer is minimized.

Acoustic quieting might consider...

- Noise generation: by limiting the noise at its source,
- Sympathetic vibrations: by acoustic decoupling,
- Resonations: by acoustic damping or changing the size of the resonator,
- Sound transmissions: by reducing transmission using many methods (depending whether the transmission is through air, liquid, or solid), or
- Sound reflections: by limiting the reflection using many methods, e.g. by using acoustic absorption (deadening) materials, trapping the sound, opening a "window" to let sound out, etc.

By analyzing the entire sequence of events, from the source to the observer, an acoustic engineer can provide many ways to quieten the machine. The challenge is to do this in a practical and inexpensive way. The engineer might focus on changing materials, using a damping material, isolating the machine, running the machine in a vacuum, or running the machine slower.

Methods of Quieting

Mechanical Acoustic Quieting

- *Sound isolation: Noise isolation is isolating noise to prevent it from transferring out of one area, using barriers like deadening materials to trap sound and vibrational energy. Example: In home and office construction, many builders place sound-control barriers (such as fiberglass batting) in walls to deaden the transmission of noise through them.*
- *Noise absorption: In architectural acoustics, unwanted sounds can be absorbed rather than reflected inside the room of an observer. This is useful for noises with no point source and when a listener needs to hear sounds only from a point source and not echo reflections. Example: In a recording studio, sound proofing is accomplished with bass traps and anechoic chambers. Wallace Sabine, an American physicist, is credited with studying sound reverberations in 1900, and Carl Eyring revised his equations in 1930 for Bell Labs. Another example is the ubiquitous use of dropped ceilings and acoustical tiles in modern office buildings with high ceilings. Submarine hulls have special coatings that absorb sound.*
- *Acoustic damping: Vibration isolation prevents vibration from transferring beyond the device into another material. Damping mounts have progressed in the industry to offer vibrational resistance in many degrees of freedom. Recent advances include shock isolators damping in at least six degrees of freedom. Acoustic damping also has uses in seismic shock protection of buildings. Motors and rotating shafts are commonly fitted with these mounts at the points where they contact the building or the chassis of a large machine.*
- *Acoustic decoupling: certain parts of a machine can be built to keep the frame, chassis, or external sh*afts from receiving unwanted vibrations from a moving part. Example: Volkswagen has registered a patent for an "acoustically decoupled underbody for a motor vehicle.". Another example: Western Digital has registered a patent for an "acoustic vibration decoupler for a disk drive pivot bearing assembly.".
- Preventing stalls: Whenever a machine undergoes an aerodynamic stall, it will abruptly vibrate.

- Preventing cavitation: When a machine is in contact with a fluid, it may be susceptible to cavitation. The sounds of gas bubbles imploding is the source of the noise. Ships and submarines which have screws that cavitate are more vulnerable to detection by sonar.
- Preventing water hammer: In hydraulics and plumbing, water hammer is a known cause for the failure of piping systems. It also generates considerable noise. A valve that abruptly opens or shuts is the most common cause for water hammer.
- Shock absorption: Just as automotive shock absorbers are used to prevent mechanical shocks from reaching the passengers in a car, they are also important for quieting shocks.
- Reduction of resonance: Essentially any piece of metal or glass has certain frequencies to which it is susceptible to resonate. A machine that resonates would make a tremendous noise. Resonance also occurs in enclosures, such as when echoes reverberate in an ocarina or the pipe of a pipe organ.
- Material selection: By choosing nonmetallic components, the transmission of sound and vibrations can be minimized. For example: instead of using rigid brass fittings, a machine using flexible plastic pipe fittings may be much quieter. In some cases air can be evacuated from a machine and sealed hermetically, the vacuum inside becoming a barrier to sound transmission.

Quieting for Specific Observers

- Underwater acoustics: All of the above types of acoustic quieting apply to submarines. Additionally, a submarine may employ a tactic that prevent sounds from reaching a listener at a particular ocean depth. Operating below the depth of the sound channel axis, where the speed of sound in water is the lowest, a submarine can prevent detection by surface ships.
- Sound refraction: Just as a submarine can use refraction to hide its acoustic signature from surface vessels, the same principle of sound refraction can be used to prevent certain observers from hearing the noise. For example, an outdoor observer close to the ground will have sound waves refracted toward him when the ground is cooler than the ambient air and away from him when the ground is hotter than the air.
- Sound Redirection: One of the obvious ways to reduce the received sound level of an observer is to place the observer out of the path of the highest amplitude sounds. For example, if we mark off a circle around a jet engine and make sound power level observations along that circle, we would expect that the sound is loudest directly in line with the jet's exhaust. Observations perpendicular to the exhaust would be significantly quieter.

- Hearing protection: An observer may be forced to wear ear plugs in areas of high ambient noise levels. This may be the only quieting method available in areas of noise pollution, such as an open-air firing range or an airport.

Electronic Quieting

- Electronic vibration control: Electronics, sensors, and computers are now employed to reduce vibration. Using high speed logic, vibrations can be damped quickly and effectively by counteracting the motion before it exceeds a certain threshold.
- Electronic noise control: Electronics, sensors, and computers are also employed to cancel noise by using phase cancellation which matches the sound amplitude with a wave of the opposite polarity. This method employs the use of an active sound generating device, such as a loudspeaker to counteract ambient noise in an area. Workers in noisy environments may favor this method over ear plugs.
- Noise reduction: In sound and video equipment, noise reduction is the process of removing noise from a signal. This is strictly for electronic noise or noise which has been detected and put into electronic form.
- Noise canceling: If both the noise and the signal are received by an electronic or digital medium, noise can be filtered from the signal electronically and retransmitted without the noise.

Hush Kit

A hush kit installed on a Swedish Air Force Sud Aviation Caravelle.

A hush kit is a device for reducing noise from an engine. It is typically used on a low-bypass turbofan engine in older commercial aircraft.

Jet aircraft generate a lot of noise, contributing to noise pollution near airports. In modern high-bypass turbofan engines, the fan mounted afore the jet engine core is very large. The bigger the fan in comparison to the jet core, the more effective the bypass air is in enveloping the jet exhaust at the rear of the engine, reducing noise. The larger turbines needed to spin the large fan slow the jet exhaust, which also reduces noise. A hush kit produces a similar effect, using several modifications to the existing engine. Primarily, a device called a multilobe exhaust mixer on the rear of the engine mixes the exhaust gases of the jet core with the surrounding air and the small amount of bypass air available. Similar systems are also employed on many modern turbofan engines as standard equipment to further reduce noise. Most kits also make further modifications to the exhaust with acoustically treated tailpipes, revised inlet nacelles and guide vanes, all of which reduce forward propagating high-pitched noise caused by the small, high-speed fan.

Modern aircraft equipped with high-bypass turbofan engines are able to comply with contemporary aviation noise abatement laws and ICAO regulations. Hush kits are used on the many older freight and passenger aircraft still in service, such as the Boeing 727 and 737-200, Douglas DC-8 and DC-9, and Tupolev Tu-154. While hushkits are normally found on older aircraft, small business jets and other aircraft too small to be fitted with large, high bypass turbofan engines will be manufactured with hushkits installed, a more economical way to meet noise restrictions than expensive engine or design changes.

Hush kits add weight to aircraft that are fitted with them. They sometimes reduce engine efficiency, which can lower aircraft range and/or cruise speed. For example, a hush kit for the Gulfstream II might weigh 234 pounds (106 kg) (total airplane weight 65,500 lb (29,700 kg)) and cause about a 1.6% reduction in aircraft range due to the hush kit's reduction in engine efficiency. Another hush kit for the Gulfstream G-II and G-III jets weighs about 170 kg (370 lb) and has a calculated 2% drop in range, although the reduction in range has not been noticed during actual flights. A different example is Federal Express's hush kit for its Boeing 727s, which adds 410 kg (900 lb) of extra weight (total airplane weight up to 86,000 kg (190,000 lb)), causing a 0.5% increase in fuel burn for short trips and no measurable increase for long flights.

While hush kits effectively reduce noise emissions from older aircraft, noise cannot always be reduced to the level of modern planes at a reasonable cost. In 1999 this has led to a dispute between the United States and the European Union. The EU's proposed noise ordinances effectively prevented the use of hush kits in Europe, reducing the value of the mostly American used airplanes so equipped and hurting the profits of American hush kit manufacturers. EU Regulation 925/99 was passed over US threats to ban Concorde, but was superseded and effectively repealed by EU Directive No. 2002/30/EC issued March 26, 2002.

Noise Barrier

A noise barrier (also called a soundwall, sound berm, sound barrier, or acoustical barrier) is an exterior structure designed to protect inhabitants of sensitive land use areas from noise pollution. Noise barriers are the most effective method of mitigating roadway, railway, and industrial noise sources – other than cessation of the source activity or use of source controls.

In the case of surface transportation noise, other methods of reducing the source noise intensity include encouraging the use of hybrid and electric vehicles, improving automobile aerodynamics and tire design, and choosing low-noise paving material. Extensive use of noise barriers began in the United States after noise regulations were introduced in the early 1970s.

History

Noise barriers have been built in the United States since the mid-twentieth century, when vehicular traffic burgeoned. In the late 1960s, analytic acoustical technology emerged to mathematically evaluate the efficacy of a noise barrier design adjacent to a specific roadway. By the 1990s, noise barriers that included use of transparent materials were being designed in Denmark and other western European countries. Below, a researcher collects data to calibrate a roadway noise model for Foothill Expressway.

Acoustical scientist measures sound in noise barrier design study, Santa Clara County, California.

The best of these early computer models considered the effects of roadway geometry, topography, vehicle volumes, vehicle speeds, truck mix, roadway surface type, and micro-meteorology. Several U.S. research groups developed variations of the computer modeling techniques: Caltrans Headquarters in Sacramento, California; the ESL Inc. group in Palo Alto, California; the Bolt, Beranek and Newman group in Cambridge, Massachusetts, and a research team at the University of Florida. Possibly the earliest published work that scientifically designed a specific noise barrier was the study for the Foothill Expressway in Los Altos, California.

Numerous case studies across the U.S. soon addressed dozens of different existing and planned highways. Most were commissioned by state highway departments and conducted by one of the four research groups mentioned above. The U.S. National Environmental Policy Act effectively mandated the quantitative analysis of noise pollution from every Federal-Aid Highway Act Project in the country, propelling noise barrier model development and application. With passage of the Noise Control Act of 1972, demand for noise barrier design soared from a host of noise regulation spinoff.

By the late 1970s, more than a dozen research groups in the U.S. were applying similar computer modeling technology and addressing at least 200 different locations for noise barriers each year. As of 2006, this technology is considered a standard in the evaluation of noise pollution from highways. The nature and accuracy of the computer models used is nearly identical to the original 1970s versions of the technology.

Theory of Design

The acoustical science of noise barrier design is based upon treating an airway or railway as a line source. The theory is based upon blockage of sound ray travel toward a particular receptor; however, diffraction of sound must be addressed. Sound waves bend (downward) when they pass an edge, such as the apex of a noise barrier. Further complicating matters is the phenomenon of refraction, the bending of sound rays in the presence of an inhomogeneous atmosphere. Wind shear and thermocline produce such inhomogeneities.

The sound sources modeled must include engine noise, tire noise, and aerodynamic noise, all of which vary by vehicle type and speed. The resulting computer model is based upon dozens of physics equations translated into thousands of lines of computer code. Software applications are available which are able to model these situations and assist in the design of such noise barriers.

Noise barrier earth berm along California State Route 12, Sonoma County, California

Some noise barriers consist of a masonry wall or earthwork, or a combination thereof (such as a wall atop an earth berm). Sound abatement walls are commonly constructed using steel, concrete, masonry, wood, plastics, insulating wool, or composites. In the most extreme cases, the entire roadway is surrounded by a noise abatement structure, or dug into a tunnel using the cut-and-cover method. The noise barrier may be constructed on private land, on a public right-of-way, or on other public land. Because sound levels are measured using a logarithmic scale, a reduction of nine decibels is equivalent to elimination of approximately 80 percent of the unwanted sound energy.

Noise barrier walls made of absorptive material mitigate sound differently than those that are a hard surface. A sound reducing barrier wall with a porous surface material and sound-dampening content material is said to be absorptive. This means little or no noise is reflected back towards the source or elsewhere. Hard surfaces such as masonry or concrete are considered to be reflective. This means most of the noise is reflected back towards the noise source and beyond.

Noise barriers can be extremely effective tools for noise pollution abatement, but theory calculates that certain locations and topographies are not suitable for use of any reasonable noise barrier. Cost and aesthetics play a role in the final choice of any noise barrier.

Tradeoffs

This noise abatement wall in The Netherlands has a transparent section at the driver's eye-level to reduce the visual impact for road users.

Disadvantages of noise barriers include:

- Aesthetic impacts for motorists and neighbors, particularly if scenic vistas are blocked
- Costs of design, construction, and maintenance
- Necessity to design custom drainage that the barrier may interrupt

Normally, the benefits of noise reduction far outweigh aesthetic impacts for residents protected from unwanted sound. These benefits include lessened sleep disturbance, improved ability to enjoy outdoor life, reduced speech interference, stress reduction, reduced risk of hearing impairment, and a reduction in the elevated blood pressure created by noise (that improves cardiovascular health).

Sound barrier walls vary in cost depending on the type and quality. Concrete is popular due to lower cost. Since they are reflective, they could potentially create noise for those across from the barrier. Absorptive barriers absorb, and thus abate the noise more effectively, but cost more as they are often custom built and are composed of a variety of materials that allow for absorption. With regard to berm construction costs, a major factor is the availability of excess soil in the immediate area which could be used for berm construction. If the soil is present, it is often cheaper to construct an earth berm noise barrier than to haul away the excess dirt, provided there is sufficient land area available for berm construction. Generally a four-to-one ratio of berm cross sectional width to height is required. Thus, for example, to build a 6-foot-high (1.8 m) berm, one needs an available width of 24 feet (7.3 m).

Effect on Downwind Pollution Concentrations

Roadside noise barriers have shown to reduce the near road air pollution concentration levels. Within 15–50 m from the roadside, air pollution concentration levels at the lee side of the noise barriers can be to reduce up to about 50% compared to open road values.

Noise barriers force the pollution plumes coming from the road to move up and over the barrier creating the effect of an elevated source and enhancing vertical dispersion of the plume. The deceleration and the deflection of the initial flow by the noise barrier, force the plume to disperse horizontally. A highly turbulent shear zone characterized by slow velocities and a re-circulation cavity is created in the lee side of the barrier and further enhances the dispersion; all of which result in a well-mixed zone with lower pollutant concentrations downwind behind the barrier.

Noise Regulation

Noise regulation includes statutes or guidelines relating to sound transmission established by national, state or provincial and municipal levels of government. After the watershed passage of the United States Noise Control Act of 1972, other local and state governments passed further regulations.

A noise regulation restricts the amount of noise, the duration of noise and the source of noise. It usually places restrictions for certain times of the day.

Although the UK and Japan enacted national laws in 1960 and 1967 respectively, these laws were not at all comprehensive or fully enforceable as to address generally rising ambient noise, enforceable numerical source limits on aircraft and motor vehicles or comprehensive directives to local government.

United States Initial Legislation

In the 1960s and earlier, few people recognized that citizens might be entitled to be protected from adverse sound level exposure. Most concerted actions consisted of citizens groups organized to oppose a specific highway or airport, and occasionally a nuisance lawsuit would arise. Things in the United States changed rapidly with passage of the National Environmental Policy Act (NEPA) in 1969 and the Noise Pollution and Abatement Act, more commonly called the Noise Control Act (NCA), in 1972. Passage of the NCA was remarkable considering the lack of historic organized citizen concern. However, the United States Environmental Protection Agency (EPA) had testified before Congress that 30 million Americans are exposed to non-occupational noise high enough to cause hearing loss and 44 million Americans live in homes impacted by aircraft or highway noise.

NEPA requires all federally funded major actions to be analyzed for all physical environmental impacts including noise pollution, and the NCA directed the EPA to promulgate regulations for a host of noise emissions. Many city ordinances prohibit sound above a threshold intensity from trespassing over property line at night, typically between 9 p.m. to 7 a.m., and during the day restricts it to a higher sound level; however, enforcement is uneven. Many municipalities do not follow up on complaints. Even where a municipality has an enforcement office, it may only be willing to issue warnings, since taking offenders to court is expensive. A notable exception to this rule is the City of Portland, Oregon, which has instituted an aggressive protection for its citizens with fines reaching as high at $5000 per infraction, with the ability to cite a responsible noise violator multiple times in a single day.

Japan

Japan actually passed the first national noise control act, but its scope was much more limited than the U.S. law, addressing mainly workplace and construction noise.

Follow-up on Initial U.S. Laws

Initially these laws had a significant effect on thoughtful study of transportation programs and also federally funded housing programs in the United States. They also gave states and cities an impetus to consider environmental noise in their planning and zoning decisions, and led to a host of statutes below the federal level. Awareness of the need for noise control was rising. In fact, by 1973 a national poll of 60,000 U.S. residents found that sixty percent of people considered street noise to have a "disturbing, harmful or dangerous" impact.

This trend continued strongly throughout the 1970s in the U.S., with about half of the states and hundreds of cities passing substantive noise control laws. Noise regulation subsided sharply in 1981, when Congress ended funding for the NCA. EPA had pre-empted lower levels of government from regulating sources, so states could not legislate standards such as for truck noise emissions. Thus, in areas where the federal government had failed to promulgate clear standards (such as aircraft noise), no further progress could be made except by the Federal Aviation Administration (FAA), which has an inherent conflict of interest regarding noise regulation.

Nevertheless, some states continued to act. California carried out an ambitious plan to require its cities to establish a "Noise Element of the General Plan," which provides guidance for land planning decisions to minimize noise impacts on the public. Many cities throughout the U.S. also have noise ordinances, which specifies the allowable sound level that can cross property lines. These ordinances can be enforced with local police powers.

Europe and Asia

Several European countries emulated the U.S. national noise control law: Netherlands (1979), France (1985), Spain (1993), and Denmark (1994). In some cases unlegislated innovations have led to quieter products exceeding legal mandates (for example, hybrid vehicles or best available technology in washing machines).

The U.S. activities lag by 10 to 20 years behind most European countries. Russia, China and undeveloped countries lag even further behind.

National Controls in the U.S. program

After the passage of the NCA, EPA promulgated regulations setting maximum noise limits on a gamut of motor vehicles, industrial machinery and household appliances. The Agency conducted extensive testing and consulted with industry on the practicality of manufacturing quieter devices. EPA's efforts had an influence on the future of a quieter generation of machines. However, roadway noise and aircraft noise account for the lion's share of noise emissions, and the EPA standards for those vehicles pre-empted states from further regulating. In the case of aircraft noise, FAA had veto power over EPA recommendations, so those standards never pushed the envelope.

In the case of motor vehicles, states could not exact a greater standard for enforcement against an individual vehicle, and interstate commerce priorities meant that guidelines for total noise exposure along federally funded highways remained guidelines rather than strict standards. Despite these drawbacks, states and the public at large had a superb weapon in the review of proposed major transportation systems in the form of NEPA and the NCA. In many cases courts were able to enforce the intent of those laws to secure the redesign of roadways and transit systems to provide more noise mitigation or to select an alternative of lesser impact than the original project; in many other cases, the highway agencies simply listened to public input and acoustical scientists before finalizing highway and transit designs.

In the case of airport expansions, courts consistently upheld the sovereignty of the FAA over the EPA, in allowing air traffic needs to be met over environmental concerns. Thus airports were required to study impacts of air traffic and facilities expansions and provide detailed noise contour maps, but in the final analysis the EPA exposure guidelines only advisory in nature. To respond to the shortcomings of the voluntary guidelines, FAA created a well funded program to insulate thousands of homes in the vicinity of major airports. The program was based upon computer modeling of alternative insulation strategies, calculated on a house-by- house basis. While this program did nothing to mitigate exterior sound levels, it benefited residential interiors significantly.

U.S. State and Local Planning

States passed two different types of legislation starting in the 1970s, echoing the federal lead in noise control. First, many states, with California in the vanguard on a state level, began requiring each municipality and county to have a Noise Element of the General Plan, a substantial noise data base and blueprint for making land use decisions in that jurisdiction. The Noise Element became an integral part of the municipal or county General Plan, especially in California. This document compiled a comprehensive set of measurements setting forth existing sound levels, frequently in the form of sound level contour maps to illustrate where varying sound levels fall relative to land use categories.

The Noise Element further states goals for each land use class and even numerical planning standards in order to evaluate future development proposals with regard to noise pollution. Technical analysis of urban highway noise had advanced by the early 1970s to allow intricate analysis of urban planning decisions in order to plan and design urban highways and support associated noise regulations.

Cities and counties in the U.S., who either fell under state mandates or who voluntarily chose to control noise through land use decisions, were active in categorizing sound levels and seeking development strategies that would minimize the number of persons exposed to harmful levels of (primarily) motor vehicle noise. Portland, Oregon continues to innovate through its almost 35-year-old Noise Control Office at the City's Bureau of Development Services. Today its code is still one of the only comprehensive codes in the U.S. that not only regulates based on a given decibel level, but also includes sound limitations based on the specific pitch or frequency of the given noise.

Local Noise Ordinances in U.S. and Europe

Local ordinances are principally aimed at construction noise, power equipment operated by individuals and unmuffled industrial noise penetrating residential areas. Thousands of U.S. cities have prepared noise ordinances that give noise control officers and police the power to investigate noise complaints and enforcement power to abate the offending noise source, through shutdowns and fines. In the 1970s and early 1980s there was even a professional association for noise enforcement officers called NANCO, "National Association of Noise Control Officials."

Today only a handful of properly trained Noise Control Officers remain in the United States. A typical noise ordinance sets forth clear definitions of acoustic nomenclature and defines categories of noise generation; then numerical standards are established, so that enforcement personnel can take the necessary steps of warnings, fines or other municipal police power to rectify unacceptable noise generation. Ordinances have achieved certain successes but they can be thorny to implement. Many European cities are still treating noise as the U.S. did in the 1960s, as a nuisance and not as a numerical standard to be achieved.

- Effects of Noise on Health and Welfare

One obligation of a community is to protect its citizens from adverse environmental influences. Noise is one of these factors, Noise has documented effects on people, they can be divided into three types. The first type is a physical effect that directly and adversely effects a person's health. Hearing loss and vibration of bodily components are examples. The second type is a physiological effect that adversely effects a person's health; heightened blood pressure and general stress response are examples. The third type is psychological that adversely effects a person's welfare; examples are distraction, annoyance, and complaint. The only feasible legal basis for a community's right to control noise is based on these adverse health and welfare effects. It is clearly easier to uphold the constitutionality of a noise ordinance in a court of law if it can be shown that it is based on health and welfare concerns. The following is a short list of recognized effects of noise that can be addressed as a reason for a noise ordinance.

> Excess non-Occupational noise exposure, hearing loss on both public and private property, speech interference on both public and private property, audio interference on both public and private property, and sleep interference on mostly private property.

Some Legal Considerations in the United States

There are several fundamental issues that shape the legality, effectiveness and enforceability of any community noise regulation.

Preemption

The federal government has preempted certain areas of noise regulation. They can be found in the Code of Federal Regulations under the EPA Noise Abatement Programs; Parts 201 to 205 and 211 cover railroads, motor carriers in interstate commerce, construction equipment, motor vehicles. They require product labeling and prohibit tampering with noise control devices. Communities may enact regulations that are no more strict than the federal ones so that local enforcement can be carried out. They can enact curfews and restrict vehicle use in established zones such as residential. Any restriction on interstate motor carriers or railroads may NOT be for the purpose of noise control.

States have police powers granted by the Constitution. They may also enact regulations that are no more strict than federal regulations. They may also preempt local ordinances. California and New Jersey have comprehensive noise codes that communities must meet. Many states required that local ordinances be no more strict than the state code whether such code exists or not. One relatively common preemption is protection of shooting ranges from noise regulation or litigation and right to farm laws that protect agricultural areas from nuisance litigation by encroaching residential areas.

Constitutional Vagueness

In one case, the United States Supreme Court has said Void for vagueness "It is a basic principle of due process (Amendment V) that an enactment is void for vagueness if its prohibitions are not clearly defined. Vague laws offend several important values. First, because we assume that man is free to steer between lawful and unlawful conduct, we insist that laws give the person of ordinary intelligence a reasonable opportunity to know what is prohibited, so that he may act accordingly. Vague laws may trap the innocent by not providing *fair warning*. Second, if arbitrary and discriminatory enforcement is to be prevented, laws must provide explicit standards for those who apply them. A vague law impermissibly delegates basic policy matters to policemen, judges, and juries for resolution on an ad hoc and subjective basis, with the attendant dangers of *arbitrary and discriminatory application*. Third, but related, where a vague statute abuts upon sensitive areas of basic First Amendment freedoms, it operates to inhibit the exercise of those freedoms. (the chilling effect). In one case, the court declared that numerical sound levels were constitutional as well as the term *plainly audible* provided it was associated with a reasonable distance. Two requirements for a noise ordinance provision is that:

1. provide fair warning
2. avoid the possibility of arbitrary enforcement

Free Speech

Examples are: slander, hate speech, and falsely calling **fire** in an assembly. Generally, the time, place, or manner of restrictions must:

1. be content neutral
2. be narrowly tailored
3. serve a significant government interest
4. leave open alternative channels of communication

One example concerned protesters outside an abortion clinic. The US Supreme Court stated that limitations placed on noise making were necessary to ensure the well-being of patients.

Overbreadth

The US Supreme Court has addressed the issue of too much regulation by defining overbreadth as: *A clear, precise enactment may nevertheless be overbroad if, in its reach, it prohibits constitutionally protected conduct.* One means to avoid being overbroad is by having provisions that are specific so that potential violators are given fair warning of what is prohibited. In one case the court ruled that the specificity of the city ordinance (Section 17-35) regulating school verbal protests was not *constitutionally vague*, gave *fair warning*, and was not an invitation to arbitrary enforcement and so was not overbroad, despite the implied limitation on free speech.

Nuisance

Nuisances are part of tort law. Torts are civil wrongs whether intentional or accidental from which injury occurs to another. Nuisances can be public or private. A public nuisance is an unreasonable interference with a right common to the general public while a private nuisance is an interference with a person's enjoyment and use of his land. Black's Law Dictionary defines a nuisance as... *that which annoys and disturbs one in possession of his property, rendering its ordinary use and occupation physically uncomfortable to him.* It defines a private nuisance as:..*anything done to the hurt or annoyance of the lands, tenements, or hereditaments of another. As distinguished from public nuisance, it includes any wrongful act which destroys or deteriorates the property of an individual or of a few persons or interferes with their lawful use or enjoyment thereof, or any act which unlawfully hinders them in the enjoyment of a common or public right and causes them a special injury different from that sustained by the general public.* The word *nuisance* is connected closely with the word *disturbance* which is used extensively in many noise ordinances. A definition below further restricts the meaning to *noise disturbance*. Nuisance law applies to both community noise regulation as well as private suits brought to court to reduce noise impact.

Enforceability

Care must be taken in writing a *subjective* noise provision so that it overcomes the objections listed above. Care must be taken when writing an *objective* noise provision to make sure that the sound levels are physically realizable. For example, requiring the maximum sound level of an automobile to be 40 dB(A) or the maximum sound level in a residential zone to be 30 dB(A) opens the provision to an enforceability challenge

Confiscation

Confiscation is the taking of private property for public use without compensation. It may occur legally when the government seizes property used in illegal practices. Confiscation may occur without an arrest of a person because it is seen as an arrest of property, rather than a person, and the necessity of finding a party guilty is not required in all cases. In many cases, the guilt or innocence of the property owner is irrelevant and the government need not prove anything *beyond a reasonable doubt*. In order to seize property, there must be either a warrant or a showing of probable cause. The mere showing of probable cause to support the forfeiture of an individual's property has been a controversial issue. Court opinion varies, but probable cause most often means that the government need only demonstrate reasonable grounds for the belief that the property is properly subject to forfeiture, which can be supported by something less than prima facie proof, but more than mere suspicion. The Peoria, Illinois noise ordinance (Sections 15-75 and 15-77) allows confiscation of vehicles that emit excessive sound caused by sound producing devices. Colorado Springs, Colorado (Section 9.8.102) permits confiscation of sound producing devices.

Four Types of Noise Regulation

Fixed sound sources must be treated differently than moving sources. In the former case, the listener is normally defined while for moving sources it is not. Historically, regulations were enforced by the subjective judgment of an enforcing officer. With the advent of sound measuring equipment, the judgment can be based on measured sound levels. Most comprehensive noise ordinances contain four types of provisions.

Subjective Emission These regulations allow an official to *decide* if the output of a sound source is acceptable *without* recourse to sound measurements and without regard to the presence of a specific listener. Regulations with *plainly audible* terms on public property as a criterion are examples.

Subjective Immission These regulations allow an official to *decide* if the sound received by a listener is acceptable *without* recourse to sound measurements and without regard for the specific sound power generated by the source.. Regulations with *plainly audible* or *noise disturbance* terms on private property as a criterion are examples.

Objective Emission These regulations require an official to *measure* the output of a sound source to determine whether it is acceptable *without* regard to the presence of a specific listener. Regulations with specific maximum sound output levels for motor vehicles are examples.

Objective Immission These regulations require an official to *measure* the sound received by a listener to determine whether it is acceptable *without* regard to the specific sound power generated by the source. Regulations with maximum allowable sound levels on property lines are examples

Some Definitions Used in the United States

many communities have definitions that are local to them, such as those defining motor vehicles and sound levels and sound level measurements. Some that have been added to make noise enforcement more specific are listed here.

Engine Braking Device

A compression braking device installed on large motor vehicles to assist in reduction, or control, of vehicle speed. When activated, the engine converts from a power source to a power absorber by converting the engine into an air compressor.

Muffler

Any device for the abatement of sound emission while permitting the transfer of gas. A muffler is considered to be in good working order if the sound reduction is equal to, or greater than, that of the original equipment.

Noise Disturbance

Any sound or vibration which:

1. may disturb or annoy reasonable persons of normal sensitivities or;
2. causes, or tends to cause, an adverse effect on the public health and welfare or;
3. endangers or injures people or;
4. endangers or injures personal or real property.

This can also be defined as *noise nuisance.*

Place of Public Entertainment

Any location, exterior, or interior, to a building that regularly permits public entrance for entertainment purposes. For this purpose, "public" means citizens of all types, including but not limited to, children, and private or public employees.

Plainly Audible Sound

Any sound for which the information content is unambiguously communicated to the listener, such as, but not limited to, understandable speech, comprehension of whether a voice is raised or normal, repetitive bass sounds, or comprehension of musical rhythms, without the aid of any listening device.

Powered Model Vehicle

Any self-propelled airborne, water-borne, or land-borne, plane, vessel, or vehicle, which is not designed to carry persons, including, but not limited to, any model airplane, boat, car, or rocket.

Real Property Boundary

An imaginary line along the ground surface, and its vertical extension, which separates the real property owned by one person from that owned by another person, but not including intra-building real property divisions.

Sound Reproduction Device

Any device, instrument, mechanism, equipment or apparatus for the amplification of any sounds from any radio, phonograph, stereo, tape player, musical instrument, television, loudspeaker or other sound-making or sound-producing device or any device or apparatus for the reproduction or amplification of the human voice or other sound.

Vibration Perception Threshold

The minimum ground or structure borne vibrational motion necessary to cause a person of normal sensitivity to be aware of the motion through contact, hearing, or through visual observation of moving objects.

Ordinance Provisions for Stationary Sources in the United States

there are three levels of regulation for stationary sound sources. The most basic is the general one associated with noise disturbance. It is a very broad *subjective immission* control that has evolved from earlier disturbance of the peace provisions. Subjectivity can lead to arbitrary enforcement. The next level of regulation is less broad; it is an *objective immission* control that uses specific levels of sound considered to be a noise disturbance. Arbitrary enforcement is reduced. In both cases, however, the person creating the sound may not be aware that his actions are in violation. The concept that a potential violator should have *fair warning* that his actions are in violation has led to provisions that address specific noise problems. The sections below list those that are found in community noise ordinances.

Air Conditioning, Heating and Pool Equipment

This provision is a *subjective immission* control. An evaluation of the noise disturbance is made at the listener without a sound level meter. This provision is mostly applied in residential zones such as homes, apartments and condominiums. Albuquerque, NM requires that such units the maximum permissible sound levels and recommend that those units be placed away from other residential units or on roof tops to diminish impact.

Airport and Airport Operations

Community control of airport created noise is limited to those sounds not related to flight operations. The community is able to control the land use around the airport however.

Animals

This provision is a *subjective immission* control. Most relate to barking dogs and put an upper time limit for continuous sound from them. New Jersey (Chapter 13:1G) considers a violation if the sound is continuous for more than 5 minutes or intermittent for more than 20 minutes. They also consider it a defense to violation if the animal is provoked to bark. Connecticut (Chapter 442)exempts animal sounds while Anchorage, AK (Chapter 15.70) requires that continual violations permit the animal to be taken and put out for adoption.

Authorized Outdoor Discharge of Firearms and Shooting Ranges

This provision contains only a curfew since most states protect shooting ranges from liability for noise disturbance. It can include a curfew requirement and a requirement for a public hearing if expansion of the range is desired. The provision may prohibit other weapons such as rocket propelled projectiles but may exempt unpowered weapons such as arrows. South Carolina (title 31 Chapter 18) requires that a sign stating SHOOTING RANGE-NOISE AREA be placed on all primary roads. Arizona (ARS 17-602) places a curfew from 10pm to 7 am. It also allows a tradeoff between the number of events and the maximum permitted sound level. New York (Chapter 150) also trades off overall levels with the duration of the sounds. Colorado declared that noise restrictions on shooting ranges is a detriment to public health, welfare, and morale!

Condominiums and Apartments

This provision is a *subjective immission* control. it It is designed to limit the noise disturbance between living units as defined by an enforcement official. One criterion used to evaluate that disturbance is use of *plainly audible* but at the location of the listener instead of at a specific distance. However, Charlotte, NC (Sec. 15-69) limits indoor levels to 55 dB(A) be tween 9am to 9pm and 5 dB less at other times, but only from sound reproducing devices. Salt Lake Valley Health Department (Chapter 4), Minneapolis, MN (Chapter 389), and Albuquerque, NM use levels to existing ambient to define a violation. Albuquerque NM and Omaha, NE (Chapter 17) require that intruding sounds not be audible. Burlington, VT (Chapter 18) requires that renters be supplied with the city noise ordinance.

Construction

This provision can be both a *subjective immission* control and an *objective emission* control. Normally there are daily curfews and in some case weekend curfews. The subjective aspect is to prevent noise disturbance in the adjacent community. The objective aspect is to control the sound output of specific machines. There are four major sources of site noise: (1) direct sound from continually operating equipment such as air compressors;

(2) intermittent sound from equipment such as jack hammers;(3)backup alarms; and (4) hauling equipment such as trucks. Air compressor noise is regulated by CFR 204 and backup alarms are regulated by CFR 1926. Boston, MA(Section 16-26.4) permits construction on weekdays between 7am and 6 pm. Madison, WI (Chapter 24.08) limits sound levels to 88 dB(A) at 50 Feet. Miami, FL (Section 36-6) considers the noise a noise disturbance if it occurs between 6pm and 8am during the week and any time on Sunday. Dallas, TX (Section 30-2 (9)) permits construction in residential zones from 7am to 7pm on weekdays, from 9am to 7pm on Saturdays and Holidays, and prohibits construction on Sundays. Albuquerque,NM (Section 9-9-8) has a more complex control. It prohibits construction and demolition within 500 feet of noise sensitive properties (residences included) if the equipment sound control devices are less effective than the original equipment and if noise mitigation measures are not used when the levels exceed 90 dB (weighting not specified) or more than 80 dB during the day for three days.

Domestic and Commercial Power Tools

This provision is a *subjective immission* control with a curfew. It is used in residential zones as well as in commercial areas abutting residential zones. Portland, OR (Title 18.10.030) has several ways to handle these tools. They separate *outdoor* and *indoor* use to different maximum levels at the property line. They have a night curfew and separate 5 HP tools from higher powered tools. Madison WI (Chapter 24.08) has similar HP restrictions. Albuquerque, NM (Title 9-9-7) restricts locations to more than 500 feet from residential and noise sensitive zones. Dallas, TX(Sec. 30-2) exempts lawn maintenance tools during daylight hours. Green Bay, WI (Subchapter II – 27.201) exempts snow removal tools.

Explosives, Firearms, Impulsive Sources, and Similar Devices

This provision is a *subjective immission* control with a curfew. It is for impulsive sound sources that are not associated with construction activities or shooting ranges. Many communities use the Maximum Permissible Sound Levels criterion, with a correction for the character of the sound. Illinois () sets maximum blasting levels by land use zone and in three time categories. Portland, OR (Section 18.10.010.F) limits levels to 100 dB (peak) from 7am to 10pm and 80 dB (peak) from 10pm to 7 am.

Fracking Operations

Hydraulic Fracturing operations generate site sound as well as vehicle sound and several different provisions are required to control it. Federal law regulates the levels of certain site machinery. A *subjective immission* control or a *objective immission* control can be applied to surrounding neighborhoods. Motor vehicle sound is mostly off-site so vehicle noise regulations are applicable. The State of New York has announced a statewide ban on such operations. Buffalo, NY and Pittsburgh, PA have announced a ban. Colorado has numerous activities to stop fracking.

Funerals

This is a *subjective emission* control to reduce the excessive shouting and protests that can surround funeral proceedings. It makes use of the *plainly audible* term and so adds a distance criterion. There are certain groups, particularly those that object to involvement in foreign wars, who believe it is an obligation to disrupt and picket funerals, especially those of deceased military veterans. This provision must not infringe on Constitutionally protected free speech. Illinois h(720 ILCS 5/26 6) has a comprehensive provision covering more aspects of this event than noise. They failed to use "plainly audible" which is narrower than "audible". Utah(Section 76-9-108) restricts disruptive activity to beyond 200 feet.

Loading and Unloading Operations

This provision is a *subjective immission* control with a curfew. Operations in commercial facilities can impact adjacent residential zones. Los Angeles, CA (Section 114.03) places a curfew on such operations between 10pm and 7am but only if the source is within 200 feet of a residence. Chicago, IL (Section 11-4-2830) permits night operations unless they create a noise disturbance. Hammond, IN (Section 6.2.6) prohibits noise disturbance between 7pm and 7am.

Maximum Permissible Sound Pressure Levels

This provision is an *objective immission* control. It requires the measurement of sound levels at or beyond a property line and its vertical extension. There are several methods for implementing such a provision:

1. It may not permit any exceedence or may permit exceedence only for a percentage of the measurement period.
2. It may require the measurement method to be instantaneous, such as dB(A) or time-averaged, such as Energy Equivalent Level (Leq).
3. It may be a fixed level limit, such as 55 dB(A), or it may be a level relative to the ambient sound, such as 5 dB(A) above the ambient.
4. It may require measurement of the frequency spectrum, such as one octave bands, or A-weighting, such as dB(A).
5. It may define different maximum levels based on zoning criteria, such as residential, commercial, or industrial.
6. It may define different maximum levels based on time-of-day or day-of-week, such as reduced maxima during night hours or on weekends.

7. It may require reduction of maximum levels based on the character of the sound, such as intermittent or impulsive.

8. It may exempt certain classes of sound sources, such as shooting ranges, farm equipment, emergency equipment, railroads, or licensed activities.

most noise ordinances set maximum levels for two time periods: Day (7am to 10pm) and Night (10pm to 7am). San Diego sets three periods: Day (7am to 7pm), Evening (7pm to 10pm), and Night (10pm to 7am) and exempts industrial zones from time based restrictions. Seattle, WA (Chapter 25.08) sets two time periods but changes 7am to 9am on weekends and holidays. Several states have maximum permissible land use sound levels in dB(A). Most have Day and Night periods and three use categories: residential, Commercial and Industrial. Washington (Chapter 70.107) sets maximum levels in dB(A) but allows 5 dB(A) more if the sound is only 15 minutes in an hour, or 10 dB(A) for 5 minutes in an hour. Numerous cities have fixed levels, permitting excess levels for short times (e.g., Dallas, TX, Chapter 30) while others use Leq (Lincoln, NE, Chapter 8.24). Los Angeles, CA (Chapter XI) uses a relative level with a stated but *presumed* ambient. New York City, NY (Chapter 19) requires Leq measurements to be made over one hour. Atlanta, GA () limits impulsive sound to 100 dB(C) at property lines, while most reduce the maximum level by 5 dB for pure tones and impulsive sounds.

Motor Vehicle or Motorboat Repairs and Testing

This provision is a *subjective immission* control with a curfew. If the activity is done in a residential zone, the Domestic Tools provision can be applied for the repairs, but this provision also is used for the testing phase of any repairs. Los Angeles, CA (Section 114.01) covers this violation in three ways. The first is application of the *noise disturbance* provision in residential districts between the hours of 8 pm to 8 am. The second is being *plainly audible* at a distance of 150 feet or more in residential districts between the hours of 8 pm to 8 am. The third is exceeding the presumed ambient by 5 dB. Hammond, IN (Section 6.2.7) prohibits this activity as a *noise disturbance* at any time.

Noise Disturbance Prohibited

This provision is a *subjective immission* control. This provision is generic in that it covers all events that are considered disturbing by a listener, with or without measurement. The strength of this provision is that it covers situations not contemplated in a noise ordinance and can be used as backup for more specific provisions. The weakness is that it may not give *fair warning*, may lead to arbitrary enforcement on the part of the regulator, or permit unreasonable demands by a listener.

Noise Sensitive Zones

This provision is a *subjective immission* control. It is used to reduce levels of both stationary and vehicular sound sources around hospitals, schools, and other noise sensitive locations. To provide *fair warning* visible signs must be posted. It is possible to have an extensive list. For example, if churches are on the list and the community has many of them, signage, compliance, and enforcement can be a problem. In modern hospital environments, helicopter sound is exempt.

Places of Public Entertainment

This provision is an *objective immision* control. It regulates the site of the sound source while the Sound Reproduction Devices section regulates the devices that create the sound. It can regulate the sound levels received by *involuntary* listeners in the surrounding community as well as the sound levels received by *voluntary* listeners. If the latter aspect is incorporated, limiting internal sound levels often resolves community noise impact. Los Angeles requires warning signs and limits noise exposure to 95 dB(A) at any position normally occupied. Seattle, WA {Section 25.08.501} considers the sound emitted to be in violation if the sound is *plainly audible* within a dwelling from 10 pm to 7am; the need for a sound level meter is avoided. Chicago, IL {Section 11-4-2805} limits received sound levels to 55 dB(A) inside a residential dwelling unit but if the ambient is greater, the limit is 65 dB(A). If outdoors, the limit is *conversational level* at 100 feet from the property line. If the building is set back 20 feet from the property line, the allowable level is 84 dB(A)! Both of Chicago's limits apply from 10 pm to 8 am. Salt Lake health Department, UT {Section 4.5.11.(vii)} sets the limit at 95 dB(A) at a position that would normally be occupied by a patron and 100 dB(A) at other positions. They also require a sign stating WARNING: SOUND LEVELS ON THIS PREMISE [sic] MAY CAUSE PERMANENT HEARING DAMAGE. HEARING PROTECTION IS AVAILABLE. Anchorage, AK (Section 15.70.060.B.12) sets maximum levels for any patron at 90 dB(A).

Powered Unmanned Vehicles or Engines

This provision is a *subjective immission* control with a curfew. It has been used to regulate the sound of model aircraft on both private and public property. It applies to airborne, water borne, and land borne, unmanned vehicles. It makes no distinction between model vehicles and full size unmanned vehicles. it applies also to the engines of those vehicles. Most regulations pertained to private unmanned vehicles, normally restricted to local open fields. The development of drones with microphones, cameras and GPS has opened the door to commercial use over wider private and public properties. Since federal preemption of drone use will likely occur, it is important for this provision to make the distinction. Salt Lake Health Department (Section 4.5.15) limits activity to 800 feet from a dwelling between 10pm and 7am, or if it cases a noise disturbance. Atlanta, GA (Section 74-136(b)) uses the *plainly audible* criterion across a residential property line, on a public property from 10pm to 7am on weekdays or from 10pm to 10am on weekends or holidays for any sound source.

Propane Cannons

This provision is a subjective immission control with a curfew. A propane cannon is used to keep animals and birds from destroying commercial crops. In large fields, many are used and fired as frequent intervals. The sound levels are equivalent to the firing of a small artillery cannon. The provision may contain requirements that limit the number of cannons permitted in a specific area and the number firings per hour for each cannon. many states have a Right to Farm Act that limits nuisance litigation. Florida() stated that it was a purpose of their act to protect reasonable agricultural activities conducted on farm land from nuisance suits. They also added a section that limited expand of operations without consideration of noise. Fairfax County, VA (Sections 105-4-4 and 108-5-1) require agricultural operation to meet maximum land use regulations and prohibit unnecessary noise. British Columbia Ministry of Agriculture (Farm Nuisance Noise document) have developed a comprehensive set of rules for cannon use.

Public Address Systems

This provision is a *subjective immission* control. It can contain a *plainly audible* term or a curfew. It is applied to commercial facilities using a sound system to *deliberately* propagate mostly speech, but also music. Most cities have provisions relevant to this subject. Lakewood, CO (Sections 9.52.06 and 09.52.160) used *plainly audible* as a regulatory tool and prohibits the sounding of bells, or chimes from 10pm to 7am. Charlotte, NC (Section 15-69(a)(4)) limits levels to 60 dB(A) at 50 feet from 9am to 9pm and 50 dB(A) at other times. Indianapolis, IN (Section 391-505) addresses broadcasts from aircraft. Connecticut (Section 22a-69-1.7) exempts bells, carillons, and chimes from religious facilities.

Sound Reproduction Devices

This provision is *subjective immission* control. It may contain a numerical level or a *plainly audible* term and a curfew. it is applied to specific sources of sound as opposed to any location at which the sound is created. It is applied primarily to *amplified* sound sources. Older provisions listed several items such as televisions, phonographs, etc. Changing the title to the above addresses the real issue and allows for novel sound production devices. Numerous communities have provisions for these devices; many use *plainly audible* as the criterion, such as Omaha, NE (Section 17-3) and Buffalo NY (Section 293-4)

Stadiums and Outdoor Music Festivals

This provision can be either a *subjective or objective immission* control with a curfew. The subjective aspect relates to noise disturbance in the local community. The objective aspect limits the acceptable sound level in the local community. Illinois (Environmental Protection Act 415.25) exempts certain stadiums and exempts festivals, parades, or street fairs. Colorado Spring, CO (Section 9.8.101) has similar exemptions, but limits the sound levels to 80 dB(A) at residential locations.

Stationary Emergency Signaling Devices

This provision is an *emission* control with a list of devices that are exempt. It can have a term that limits the time periods in which emergency alarms may be tested. It can have a term that limits the activation time of burglar or fire alarms. Chicago, IL (Section 11-4-2815) limits the time for tests to 4 minutes between 9am and 5pm. Oregon (Chapter 467) prohibits sound when an emergency vehicle is stationary.

Stationary non-Emergency non-Safety Signaling Devices

This provision is an *emission* control that limits the activation period of alarms and restricts activation to a specific time-of-day or day-of-week. Los Angeles, CA () prohibits the sounding if the signal can be heard at 200 feet or more. Chicago, IL (Section 11-4-2820) considers the sound to be a noise disturbance in residential areas if the sound exceeds 5 minutes in any hour; steam whistles are exempt. Albuquerque, NM (Section 9-9-12) restricts levels to 5 dB over the ambient at a property line and applies maximum permissible residential level as well as *plainly audible* restrictions at night.

Street Sales

This provision is a *subjective immission* control with a curfew. Boston, MA (Section 16-2.2) prohibits street sales near schools or churches if there is a "disturbance of the peace". Hammond, IN (Section 6.2.4) places a curfew between 6 pm and 9 am.

Tampering

This provision prevents the modification of muffling devices that increase the emitted sound. It can also be used to prevent the commercial sale of such mufflers. Most states and communities have prohibitions on tampering with noise reduction devices. whether stationary or moving. Salt Lake County health Department (Section 4.5.10) prohibits modifications of mufflers that increase sound levels and prohibits tampering with noise rating labels.

Vibration

This provision is a *subjective emission* control. Noise disturbance caused by vibration comes in three forms. One is contact with vibrating surfaces, the second in auditory, and the third is the observation of surrounding objects movement. Objectively regulating vibration is difficult so this provision makes use of the Vibration Perception Threshold. Railroad caused vibration is preempted by federal law (CFR 201). Chicago, IL (Section 11-4-2910) uses the perception threshold method. Dallas, TX requires measurement of low frequency vibration. Maryland uses the definition of *noise* to include sound and vibration at sub-audible frequencies.

Wind Turbines

The sound created by wind turbines is caused by the blade rotation similar to aircraft propellers. Because the rotation rate is low, the frequency is also low, but the large size of many can result in disturbing sound levels, particularly in high wind areas. Most local control is done by advantageous site planning. New Hampshire (Title LXIV, Section 674:63) sets a sound level limit of 55 dB when measured at the site property line, allowing for exceptional events, such as storms. Studies have declared that wind noise can have a negative effect on health.

Ordinance Provisions for Moving Vehicles in the United States

Stationary sources have fixed positions, so it is possible to define the listeners and therefore *immission* controls are appropriate. Motor vehicles are moving sources so it is not possible to define any specific listeners so *emission* controls are appropriate. There are exceptions to this distinction. Construction equipment, and some recreation vehicles, move within a bounded area and can be considered to be time varying fixed sources. Standing motor vehicles can radiate sufficient sound to create noise disturbance. These must be treated by specific provisions.

Adequate Mufflers or Sound Dissipation Devices

This provision is an *objective emission* control. Unlike the Tampering provision, this is specific to motor vehicles. It requires that a vehicle muffler not create more sound than the original equipment which has been measured. It prohibits any modification or replacement that increases the sound emission beyond that of the original equipment. It prohibits the sale of mufflers that do not meet original equipment standards. Many states have requirements that a muffler *shall be in good working order* which is not specific enough. California (Section 27150.1)) requires that a retail seller that sells a product in violation of the muffler regulation must install a replacement muffler that meets the regulation and must reimburse the purchaser for the expense of replacement.

Airboats and Hovercraft

This provision is both an "objective emission *control and a* subjective immission *control. It can set a maximum sound level at a specified distance (typically 50 feet). It can have a curfew. It can be based on* noise disturbance *in the community. It can also require the use of ear protectors on passengers. Unlike motorboats, the sound generators on these vehicles are airborne, resulting in more noise impact. Florida (Section 327.65) requires a maximum level of 90 DB(A). Maine (Title 12 Section 13068-A)has three levels: operating, operating test and stationary test*

Engine Braking Devices

This provision is an *emission* control. It can restrict brake use for only safety purposes and by defining restricted areas. It can require that mufflers be maintained to keep emitted sound to that of the original equipment. Common terminology is Jake brakes after the Jacobs Company. Milwaukee, Wi (Section 80-69) prohibits use within city limits. Portland, OR (Section 18.10.020.B.3) prohibits use within 200 feet of a residence. Albuquerque, NM posts signs requiring proper mufflers.

Motorboats

This provision is both an *objective emission* control and a *subjective immission* control. Because they are moving sources, *objective* controls are appropriate for measurements on open waterways. Many motorboats operate in bounding areas, such as small lakes or canals, with adjacent residential areas. In this case, *immission* controls are appropriate. California (Section 654.05), Portland OR (Section 18.10.040), and Seattle, WA (Section 25.08.485) require *immission* measurements to be made at the shoreline. Many states require *emission* measurements to be made at 50 feet.

Motor Vehicle Horns or Signaling Devices

This provision is both a prohibition and an *emission* control. it limits to use for safety warning only.. It limits the sound level to a specific level at a prescribed distance. This provision is intended to limit horn use to safety and to limit the use of excessively loud air horns or *Rumbler* or *Howler* horns. California prohibits a person operating a motor vehicle to wear a headset or earplugs on both ears. . Oregon (Section 820.370) prohibits signaling sound when an emergency vehicle is stationary or returning from an emergency.

Motor Vehicles on a Public Right-of-Way

This provision is an *objective emission* control. It applies maximum sound levels to various categories of moving vehicles and for several vehicle speeds. It is the backbone of vehicle sound emission regulations. It generally requires a measurement of A-weighted sound level of a moving vehicle at a specific distance from the vehicle path (normally 50 feet). This provision has level restrictions on trucks over 10,000 GVW used locally and in interstate commerce. It also covers motorcycles of two horsepower ratings, mopeds, and all other vehicles on public rights-of-way. The federal government has set maximum levels for heavy trucks used in interstate commerce (40 CFR 202)and for motorcycles (40 CFR 205). Most states and many cities have maximum limits and they generally agree with federal standards where they apply. The most common speed division is 35 mph.

Motor Vehicle Racing Events

This provision is an *objective emission* control. It can define the method of vehicle operation that is used to define the maximum permitted sound level. It can have a curfew. Some states exempt motor vehicle racing events from noise disturbance litigation or prosecution. Arizona (Section 28-955.03) exempts racing motorcycles from maximum sound levels and muffler requirements. Illinois {Section 35.903} had detailed regulations on racing vehicles It required a 14 dB reduction in sound output, limited sound output at half mater to 115 dB(A), and no more than 105 dB(A) at 50 feet.

Motor Vehicle Sound Systems

This provision is both an *objective emission* control, a *subjective emission* control, and a *subjective immission* for vehicles on a public right-of-way. The first part limits the system sound level at a fixed distance. The second part uses the *plainly audible* definition for limiting the sound output. The third part uses the *noise disturbance* definition to limit the impact on neighboring properties and can be applied within public transportation. The most restrictive application of the plainly audible laws says that the sound cannot be audible to anyone other than the vehicle occupants. There are numerous state and community restrictions on vehicle sound systems. Louisiana prohibits the system from emitting sound outside of a vehicle. Richmond, CA also prohibits the sound from being audible outside the vehicle. Oregon prohibits sound systems being *plainly audible* at 50 feet. California prohibits sound systems that can be *heard* at 50 feet. Colorado Springs, CO, requires a measurement at 25 feet beyond the private property line or 25 feet from the source on public property; it does not specify a limiting level. In Lakewood, CO it must not be *plainly audible* beyond 25 feet. In Los Angeles, CA, it cannot be audible beyond 200 feet. In Seattle, WA, it must not be *plainly audible* at 75 feet. Chicago restricts levels to less than *clearly audible* at 75 feet. Minneapolis, MN restricts levels to less than audible at 50 feet. Albuquerque, NM restricts *plainly audible* to 25 feet, but also applies their land use limits. Cincinnati, OH restricts *plainly audible* to 50 feet. Dallas, TX prohibits sound or vibration that is detectable at 30 feet, or that violates the land use regulations. Houston, TX applies land use restrictions. Omaha, NE states the sound must not be audible at 100 feet. Hammond, IN restricts *plainly audible* to 25 feet. New Jersey states the sound must not be *plainly audible* at 50 feet between 8 am and 10 pm and not *plainly audible* at 25 feet between 10 pm and 8 am. Florida states the sound must not be *plainly audible* at 25 feet, but exempts business and political systems. Oregon and Tennessee state that the sound must not be *plainly audible* beyond 50 feet, as does Fairbanks, AK. Rhode Island specifically addresses low frequency sound that can be heard 20 feet from a closed vehicle or 100 feet otherwise. Salt Lake County Health Department, UT considers the sound a violation if it is *plainly audible* on a common carrier. Austin, TX states it must not be audible at 30 feet.

Motor Vehicle Theft Alarms

This provision is a *subjective emission* control with only an operational time limit. Los

Angeles, CA () requires silencing in 5 minutes. New York City, NY (Section 24.221(d)) requires automatic shut-off after 10 minutes and a prominent display of the local precinct number and telephone number. Boston, MA (Section 16-26.2) considers it a violation if the alarm is plainly audible at 200 feet and is on more than 5 minutes. Other states an communities have automatic shutoff times from 10 to 15 minutes. Some communities have banned such alarms.

Motor Vehicle Tire Squeal and Street Drag Racing

This provision is a *subjective immission* control. It is based on the *noise disturbance* from drag racing and tire squealing on public rights-of-way. Illinois (625 ILCS 5/11-505) prohibits such activities. Hammond, IN (Section 6.2.14) prohibits such activity if it creates a *noise disturbance.*

Railroads

Railroad activity is subject to federal regulations. Most communities do not attempt to regulate train sound. The level of train horns permitted by the Federal Railroad Administration sufficiently high that community impact occurs. One method to alleviate this sound is to have a community establish a quiet zone where the rail crossings meet federal safety standards so that horn use is not needed.

Recreational Off-road Vehicles

This provision is both an *objective emission* control and a *subjective immission* control with a curfew. It can apply to both public and private properties. Since these vehicles can move in large open areas, an *objective* control, limiting maximum sound levels at a fixed distance, is appropriate. Since they also can move in bounded areas near residences, a *subjective* control is appropriate. Numerous states and cities have *emission* controls measured at 50 feet; the most common level is 82 dB(A), which is similar to that for motor vehicles on public rights-of-way. Colorado Springs, CO (Section 9.8.204.C) requires a minimum distance of 660 feet from residences. Portland, OR (Section 18.10.020.C) requires the area must be designated for recreational vehicle use. Salt Lake health Department, UT (Section 4.5.10(x)) requires off-highway vehicles to be at least 800 feet from a dwelling during the day and has a curfew from 10 pm to 7 am. They prohibit any *noise disturbance* and require sound levels to the less than 96 dB(A) at 50 feet.

Recreational Snowmobiles

This provision is both an *objective emission* control and a *subjective immission* control with a curfew. It can apply to both public and private properties. Since these vehicles can move in large open areas, an *objective* control, limiting maximum sound levels at a fixed distance is appropriate. Since they also can move in bounded areas near residences, a *subjective* control is appropriate. Numerous states and communities have *objective* controls; the most common maximum level is 78 dB(A). Federal law (36 CFR 2.18) regulates snowmobiles on federal property at 78 dB(A), so states and communities are free to regulate snowmobile sound levels on their property. Lincoln NE (Section 8.24.110) limits levels to 78 dB(A). Maine (Section 13112, Chapter 937) exempts snowmobiles at sanctioned racing events. Illinois (625 ILCS 40 Sec. 4-4) does also.

Refuse Collection Vehicles

This provision is an *objective emission* control and a *subjective* control. It can place a maximum sound level at a specific distance for the loudest operation. It can set a curfew, or it can be based on *noise disturbance* in residential zones. Los Angeles, CA (Section 113.01) has a time limit that applies only within 200 feet of any residential building. Chicago, IL (Section 11-4-2900) considers any noise it a noise disturbance if the activity occurs between 8pm to 8am. Salt Lake City, UT (Section 4.5.6) considers it a noise disturbance if the activities occur between 10 pm to 7 am and closer than 800 feet from a dwelling. Atlanta, GA (Section 74-137(a)(5)prohibits collection between 9pm and am on a weekend day or legal holiday, except by permit. In Maryland, refuse collection is exempt during daytime hours and must meet maximum land use levels [55 dB(A)] in residential zones at night.

Standing Motor Vehicles

This provision is a *subjective immission* control. It sets a time limit on engine activity. It can also place a curfew on any engine activity. e. In Salt Lake City, UT (Section 4.5.10(xi)) it is considered a *noise disturbance* if the operation lasts more than 15 minutes. Dallas, TX (Section 30-3.1) applies the code to vehicles over 14,000 GVWR; they must be more than 300 feet from a residential zone and there is a 10-minute maximum. They also provide a list of idling vehicles that are exempt from prosecution such as buses or active concrete trucks. Hammond, IN (Section 6.2.10) limits operation to 3 minutes in an hour for vehicles over 14,000 GVWR in either public or private property. It exempts buses and taxis. Massachusetts allows idling no more than 5 minutes.

Building Codes

In the case of construction of new (or remodeled) apartments, condominiums, hospitals and hotels, many U.S. states and cities have stringent building codes with requirements of acoustical analysis, in order to protect building occupants from exterior noise sources and sound generated within the building itself. With regard to exterior noise, the codes usually require measurement of the exterior acoustic environment in order to determine the performance standard required for exterior building skin design.

The architect can work with the acoustical scientist to arrive at the best cost-effective means of creating a quiet interior (normally 45 dB). The most important elements of design of the building skin are usually: glazing (glass thickness, double pane design, etc.), roof material, caulking standards, chimney baffles, exterior door design, mail slots, attic ventilation ports and mounting of through the wall air conditioners. A special case of building skin design arises in the case of aircraft noise, where the FAA has funded extensive work in residential retrofit.

Regarding sound generated inside the building, there are two principal types of transmission. First, airborne sound travels through walls or floor/ceiling assemblies and can emanate from either human activities in adjacent living spaces or from mechanical noise within the building systems. Human activities might include voice, amplified sound systems or animal noise. Mechanical systems are elevator systems, boilers, refrigeration or air conditioning systems, generators and trash compactors. Since many of these sounds are inherently loud, the principle of regulation is to require the wall or ceiling assembly to meet certain performance standards (typically Sound Transmission Class of 50), which allows considerable attenuation of the sound level reaching occupants.

The second type of interior sound is called Impact Insulation Class (IIC) transmission. This effect arises not from airborne transmission, but rather from transmission of sound through the building itself. The most common perception of IIC noise is from footfall of occupants in living spaces above. This type of noise is somewhat more difficult to abate, but consideration must be given to isolating the floor assembly above or hanging the lower ceiling on resilient channel. Commonly a performance standard of IIC equal to 50 is specified in building codes. California has generally led the U.S. in widespread application of building code requirements for sound transmission; accordingly, the level of protection for building occupants has increased markedly in the last several decades.

U.S. Occupational Safety Regulations

The U.S. Occupational Safety and Health Administration has established maximum noise levels for occupational exposure, beyond which mitigation measures or personal protective equipment is required. In recent years, Buy Quiet programs and initiatives have arisen in an effort to combat occupational noise exposures. These programs promote the purchase of quieter tools and equipment and encourage manufacturers to design quieter equipment.

Harmonic Oscillator

In classical mechanics, a harmonic oscillator is a system that, when displaced from its equilibrium position, experiences a restoring force, *F*, proportional to the displacement, *x*:

$$\vec{F} = -k\vec{x}$$

where k is a positive constant.

If F is the only force acting on the system, the system is called a simple harmonic oscillator, and it undergoes simple harmonic motion: sinusoidal oscillations about the equilibrium point, with a constant amplitude and a constant frequency (which does not depend on the amplitude).

If a frictional force (damping) proportional to the velocity is also present, the harmonic oscillator is described as a damped oscillator. Depending on the friction coefficient, the system can:

- Oscillate with a frequency lower than in the non-damped case, and an amplitude decreasing with time (*underdamped* oscillator).
- Decay to the equilibrium position, without oscillations (*overdamped* oscillator).

The boundary solution between an underdamped oscillator and an overdamped oscillator occurs at a particular value of the friction coefficient and is called "critically damped."

If an external time dependent force is present, the harmonic oscillator is described as a driven oscillator.

Mechanical examples include pendulums (with small angles of displacement), masses connected to springs, and acoustical systems. Other analogous systems include electrical harmonic oscillators such as RLC circuits. The harmonic oscillator model is very important in physics, because any mass subject to a force in stable equilibrium acts as a harmonic oscillator for small vibrations. Harmonic oscillators occur widely in nature and are exploited in many manmade devices, such as clocks and radio circuits. They are the source of virtually all sinusoidal vibrations and waves.

Simple Harmonic Oscillator

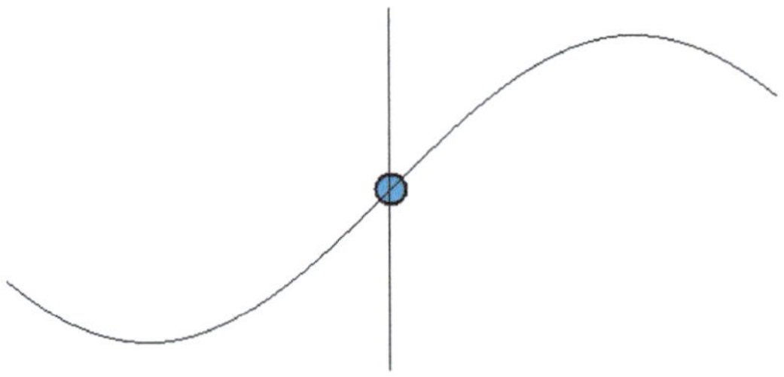

Simple harmonic motion

A simple harmonic oscillator is an oscillator that is neither driven nor damped. It consists of a mass m, which experiences a single force, F, which pulls the mass in the direction of the point x=0 and depends only on the mass's position x and a constant k. Balance of forces (Newton's second law) for the system is

$$F = ma = m\frac{d^2x}{dt^2} = m\ddot{x} = -kx.$$

Solving this differential equation, we find that the motion is described by the function

$$x(t) = A\cos\left(\omega t + \phi\right),$$

where

$$\omega = \sqrt{\frac{k}{m}} = \frac{2\pi}{T}.$$

The motion is periodic, repeating itself in a sinusoidal fashion with constant amplitude, A. In addition to its amplitude, the motion of a simple harmonic oscillator is characterized by its period T, the time for a single oscillation or its frequency $f = {}^{1}/_{T}$, the number of cycles per unit time. The position at a given time t also depends on the phase, φ, which determines the starting point on the sine wave. The period and frequency are determined by the size of the mass m and the force constant k, while the amplitude and phase are determined by the starting position and velocity.

The velocity and acceleration of a simple harmonic oscillator oscillate with the same frequency as the position but with shifted phases. The velocity is maximum for zero displacement, while the acceleration is in the opposite direction as the displacement.

The potential energy stored in a simple harmonic oscillator at position x is

$$U = \frac{1}{2}kx^2.$$

Damped Harmonic Oscillator

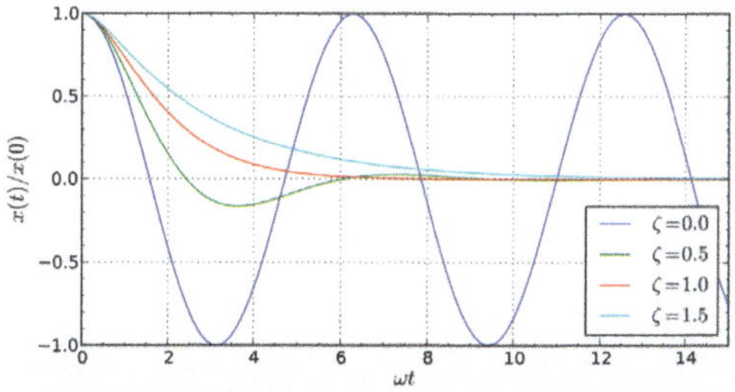

Dependence of the system behavior on the value of the damping ratio ζ

Another damped harmonic oscillator

In real oscillators, friction, or damping, slows the motion of the system. Due to frictional force, the velocity decreases in proportion to the acting frictional force. While simple harmonic motion oscillates with only the restoring force acting on the system, damped harmonic motion experiences friction. In many vibrating systems the frictional force F_f can be modeled as being proportional to the velocity v of the object: $F_f = -cv$, where c is called the *viscous damping coefficient*.

Balance of forces (Newton's second law) for damped harmonic oscillators is then

$$F = F_{ext} - kx - c\frac{dx}{dt} = m\frac{d^2x}{dt^2}.$$

When no external forces are present (i.e. when $F_{ext} = 0$), this can be rewritten into the form

$$\frac{d^2x}{dt^2} + 2\zeta\omega_0\frac{dx}{dt} + \omega_0^2 x = 0,$$

where

$\omega_0 = \sqrt{\frac{k}{m}}$ is called the 'undamped angular frequency of the oscillator' and

$\zeta = \frac{c}{2\sqrt{mk}}$ is called the 'damping ratio'.

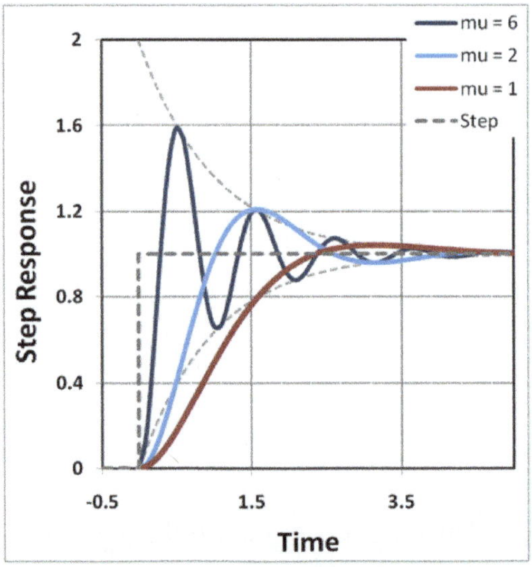

Step-response of a damped harmonic oscillator; curves are plotted for three values of $\mu = \omega_1 = \omega_0\sqrt{1-\zeta^2}$. Time is in units of the decay time $\tau = 1/(\zeta\omega_0)$.

The value of the damping ratio ζ critically determines the behavior of the system. A damped harmonic oscillator can be:

- *Overdamped* ($\zeta > 1$): The system returns (exponentially decays) to steady state without oscillating. Larger values of the damping ratio ζ return to equilibrium slower.

- *Critically damped* ($\zeta = 1$): The system returns to steady state as quickly as possible without oscillating (although overshoot can occur). This is often desired for the damping of systems such as doors.
- *Underdamped* ($\zeta < 1$): The system oscillates (with a slightly different frequency than the undamped case) with the amplitude gradually decreasing to zero. The angular frequency of the underdamped harmonic oscillator is given by $\omega_1 = \omega_0\sqrt{1-\zeta^2}$, the exponential decay of the underdamped harmonic oscillator is given by $\lambda = \omega_0\zeta$.

The Q factor of a damped oscillator is defined as

$$Q = 2\pi \times \frac{\text{Energy stored}}{\text{Energy lost per cycle}}.$$

Q is related to the damping ratio by the equation $Q = \dfrac{1}{2\zeta}$.

Driven Harmonic Oscillators

Driven harmonic oscillators are damped oscillators further affected by an externally applied force $F(t)$.

Newton's second law takes the form

$$F(t) - kx - c\frac{\mathrm{d}x}{\mathrm{d}t} = m\frac{\mathrm{d}^2x}{\mathrm{d}t^2}.$$

It is usually rewritten into the form

$$\frac{\mathrm{d}^2x}{\mathrm{d}t^2} + 2\zeta\omega_0\frac{\mathrm{d}x}{\mathrm{d}t} + \omega_0^2 x = \frac{F(t)}{m}.$$

This equation can be solved exactly for any driving force, using the solutions $z(t)$ which satisfy the unforced equation:

$$\frac{\mathrm{d}^2z}{\mathrm{d}t^2} + 2\zeta\omega_0\frac{\mathrm{d}z}{\mathrm{d}t} + \omega_0^2 z = 0,$$

and which can be expressed as damped sinusoidal oscillations,

$$z(t) = Ae^{-\zeta\omega_0 t}\sin\left(\sqrt{1-\zeta^2}\,\omega_0 t + \varphi\right),$$

in the case where $\zeta \leq 1$. The amplitude A and phase φ determine the behavior needed to match the initial conditions.

Step Input

In the case $\zeta < 1$ and a unit step input with $x(0) = 0$:

$$\frac{F(t)}{m} = \begin{cases} \omega_0^2 & t \geq 0 \\ 0 & t < 0 \end{cases}$$

the solution is:

$$x(t) = 1 - e^{-\zeta\omega_0 t} \frac{\sin\left(\sqrt{1-\zeta^2}\ \omega_0 t + \varphi\right)}{\sin(\varphi)},$$

with phase φ given by

$$\cos\varphi = \zeta.$$

The time an oscillator needs to adapt to changed external conditions is of the order $\tau = 1/(\zeta\omega_0)$. In physics, the adaptation is called relaxation, and τ is called the relaxation time.

In electrical engineering, a multiple of τ is called the *settling time*, i.e. the time necessary to ensure the signal is within a fixed departure from final value, typically within 10%. The term *overshoot* refers to the extent the maximum response exceeds final value, and *undershoot* refers to the extent the response falls below final value for times following the maximum response.

Sinusoidal Driving Force

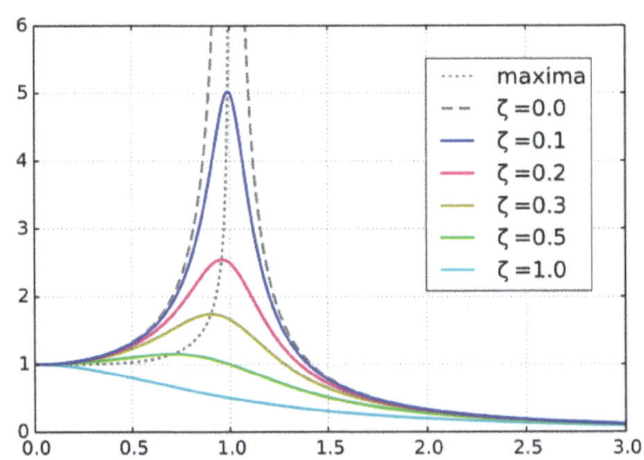

Steady state variation of amplitude with relative frequency ω / ω_0 and damping ζ of a driven simple harmonic oscillator.

In the case of a sinusoidal driving force:

$$\frac{d^2x}{dt^2} + 2\zeta\omega_0 \frac{dx}{dt} + \omega_0^2 x = \frac{1}{m} F_0 \sin(\omega t),$$

where F_0 is the driving amplitude and ω is the driving frequency for a sinusoidal driving mechanism. This type of system appears in AC driven RLC circuits (resistor-inductor-capacitor) and driven spring systems having internal mechanical resistance or external air resistance.

The general solution is a sum of a transient solution that depends on initial conditions, and a steady state that is independent of initial conditions and depends only on the driving amplitude F_0, driving frequency, ω, undamped angular frequency ω_0, and the damping ratio ζ.

The steady-state solution is proportional to the driving force with an induced phase change of ϕ:

$$x(t) = \frac{F_0}{mZ_m\omega}\sin(\omega t + \phi)$$

where

$$Z_m = \sqrt{(2\omega_0\zeta)^2 + \frac{1}{\omega^2}\left(\omega_0^2 - \omega^2\right)^2}$$

is the absolute value of the impedance or linear response function and

$$\phi = \arctan\left(\frac{2\omega\omega_0\zeta}{\omega^2 - \omega_0^2}\right)$$

is the phase of the oscillation relative to the driving force, if the arctan value is taken to be between -180 degrees and 0 (that is, it represents a phase lag, for both positive and negative values of the arctan's argument).

For a particular driving frequency called the resonance, or resonant frequency $\omega_r = \omega_0\sqrt{1-2\zeta^2}$, the amplitude (for a given F_0) is maximum. This resonance effect only occurs when $\zeta < 1/\sqrt{2}$, i.e. for significantly underdamped systems. For strongly underdamped systems the value of the amplitude can become quite large near the resonance frequency.

The transient solutions are the same as the unforced ($F_0 = 0$) damped harmonic oscillator and represent the systems response to other events that occurred previously. The transient solutions typically die out rapidly enough that they can be ignored.

Parametric Oscillators

A parametric oscillator is a driven harmonic oscillator in which the drive energy is provided by varying the parameters of the oscillator, such as the damping or restoring force. A familiar example of parametric oscillation is "pumping" on a playground swing. A person on a moving swing can increase the amplitude of the swing's oscillations without any external drive force (pushes) being applied, by changing the moment of inertia of the swing by rocking back and forth ("pumping") or alternately standing and squatting, in rhythm with the swing's oscillations. The varying of the parameters drives the system. Examples of parameters that may be varied are its resonance frequency ω and damping β.

Parametric oscillators are used in many applications. The classical varactor parametric oscillator oscillates when the diode's capacitance is varied periodically. The circuit that varies the diode's capacitance is called the "pump" or "driver". In microwave electronics, waveguide/YAG based parametric oscillators operate in the same fashion. The designer varies a parameter periodically to induce oscillations.

Parametric oscillators have been developed as low-noise amplifiers, especially in the radio and microwave frequency range. Thermal noise is minimal, since a reactance (not a resistance) is varied. Another common use is frequency conversion, e.g., conversion from audio to radio frequencies. For example, the Optical parametric oscillator converts an input laser wave into two output waves of lower frequency (ω_s, ω_i).

Parametric resonance occurs in a mechanical system when a system is parametrically excited and oscillates at one of its resonant frequencies. Parametric excitation differs from forcing, since the action appears as a time varying modification on a system parameter. This effect is different from regular resonance because it exhibits the instability phenomenon.

Universal Oscillator Equation

The equation

$$\frac{d^2q}{d\tau^2}+2\zeta\frac{dq}{d\tau}+q=0$$

is known as the universal oscillator equation since all second order linear oscillatory systems can be reduced to this form. This is done through nondimensionalization.

If the forcing function is $f(t) = \cos(\omega t) = \cos(\omega t_c \tau) = \cos(\omega\tau)$, where $\omega = \omega t_c$, the equation becomes

$$\frac{d^2q}{d\tau^2}+2\zeta\frac{dq}{d\tau}+q=\cos(\omega\tau).$$

The solution to this differential equation contains two parts, the "transient" and the "steady state".

Transient Solution

The solution based on solving the ordinary differential equation is for arbitrary constants c_1 and c_2

$$q_t(\tau)=\begin{cases} e^{-\zeta\tau}\left(c_1 e^{\tau\sqrt{\zeta^2-1}}+c_2 e^{-\tau\sqrt{\zeta^2-1}}\right) & \zeta>1 \text{ (overdamping)} \\ e^{-\zeta\tau}(c_1+c_2\tau)=e^{-\tau}(c_1+c_2\tau) & \zeta=1 \text{ (critical damping)} \\ e^{-\zeta\tau}\left[c_1\cos\left(\sqrt{1-\zeta^2}\tau\right)+c_2\sin\left(\sqrt{1-\zeta^2}\tau\right)\right] & \zeta<1 \text{(underdamping)} \end{cases}$$

The transient solution is independent of the forcing function.

Steady-state solution

Apply the "complex variables method" by solving the auxiliary equation below and then finding the real part of its solution:

$$\frac{d^2q}{d\tau^2}+2\zeta\frac{dq}{d\tau}+q=\cos(\omega\tau)+i\sin(\omega\tau)=e^{i\omega\tau}.$$

Supposing the solution is of the form

$$q_s(\tau)=Ae^{i(\omega\tau+\phi)}.$$

Its derivatives from zero to 2nd order are

$$q_s=Ae^{i(\omega\tau+\phi)},\ \frac{dq_s}{d\tau}=i\omega Ae^{i(\omega\tau+\phi)},\ \frac{d^2q_s}{d\tau^2}=-\omega^2Ae^{i(\omega\tau+\phi)}.$$

Substituting these quantities into the differential equation gives

$$-\omega^2Ae^{i(\omega\tau+\phi)}+2\zeta i\omega Ae^{i(\omega\tau+\phi)}+Ae^{i(\omega\tau+\phi)}=(-\omega^2A+2\zeta i\omega A+A)e^{i(\omega\tau+\phi)}=e^{i\omega\tau}.$$

Dividing by the exponential term on the left results in

$$-\omega^2A+2\zeta i\omega A+A=e^{-i\phi}=\cos\phi-i\sin\phi.$$

Equating the real and imaginary parts results in two independent equations

$$A(1-\omega^2)=\cos\phi \qquad 2\zeta\omega A=-\sin\phi.$$

Amplitude Part

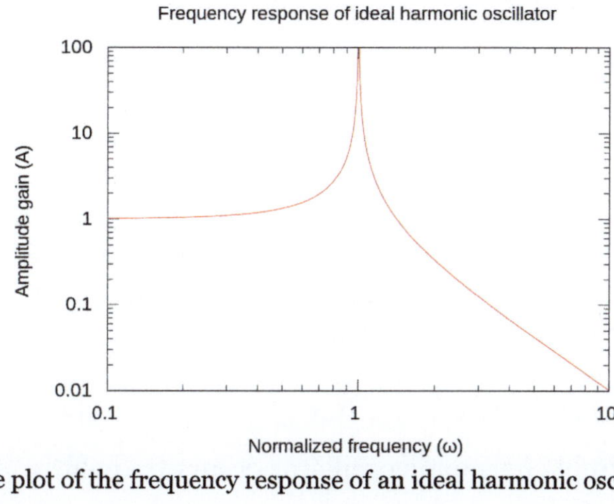

Bode plot of the frequency response of an ideal harmonic oscillator.

Squaring both equations and adding them together gives

$$\left.\begin{array}{rcl} A^2(1-\omega^2)^2 & = & \cos^2\phi \\ [6pt](2\zeta\omega A)^2 & = & \sin^2\phi \end{array}\right\} \Rightarrow A^2[(1-\omega^2)^2+(2\zeta\omega)^2]=1.$$

Therefore,

$$A = A(\zeta,\omega) = \operatorname{sign}\left(\frac{-\sin\phi}{2\zeta\omega}\right)\frac{1}{\sqrt{(1-\omega^2)^2+(2\zeta\omega)^2}}.$$

Compare this result with the theory section on resonance, as well as the "magnitude part" of the RLC circuit. This amplitude function is particularly important in the analysis and understanding of the frequency response of second-order systems.

Phase Part

To solve for φ, divide both equations to get

$$\tan\phi = -\frac{2\zeta\omega}{1-\omega^2} = \frac{2\zeta\omega}{\omega^2-1} \Rightarrow \phi \equiv \phi(\zeta,\omega) = \arctan\left(\frac{2\zeta\omega}{\omega^2-1}\right).$$

This phase function is particularly important in the analysis and understanding of the frequency response of second-order systems.

Full Solution

Combining the amplitude and phase portions results in the steady-state solution

$$q_s(\tau) = A(\zeta,\omega)\cos(\omega\tau+\phi(\zeta,\omega)) = A\cos(\omega\tau+\phi).$$

The solution of original universal oscillator equation is a superposition (sum) of the transient and steady-state solutions

$$q(\tau) = q_t(\tau) + q_s(\tau).$$

Equivalent Systems

Harmonic oscillators occurring in a number of areas of engineering are equivalent in the sense that their mathematical models are identical. Below is a table showing analogous quantities in four harmonic oscillator systems in mechanics and electronics. If analogous parameters on the same line in the table are given numerically equal values, the behavior of the oscillators—their output waveform, resonant frequency, damping factor, etc.—are the same.

Application to a Conservative Force

The problem of the simple harmonic oscillator occurs frequently in physics, because a mass at equilibrium under the influence of any conservative force, in the limit of small motions, behaves as a simple harmonic oscillator.

A conservative force is one that has a potential energy function. The potential energy function of a harmonic oscillator is:

$$V(x)=\frac{1}{2}kx^2$$

Given an arbitrary potential energy function $V(x)$, one can do a Taylor expansion in terms of x around an energy minimum ($x=x_0$) to model the behavior of small perturbations from equilibrium.

$$V(x)=V(x_0)+(x-x_0)V'(x_0)+\frac{1}{2}(x-x_0)^2V^{(2)}(x_0)+O(x-x_0)^3$$

Because $V(x_0)$ is a minimum, the first derivative evaluated at x_0 must be zero, so the linear term drops out:

$$V(x)=V(x_0)+\frac{1}{2}(x-x_0)^2V^{(2)}(x_0)+O(x-x_0)^3$$

The constant term $V(x_0)$ is arbitrary and thus may be dropped, and a coordinate transformation allows the form of the simple harmonic oscillator to be retrieved:

$$V(x)\approx\frac{1}{2}x^2V^{(2)}(0)=\frac{1}{2}kx^2$$

Thus, given an arbitrary potential energy function $V(x)$ with a non-vanishing second derivative, one can use the solution to the simple harmonic oscillator to provide an approximate solution for small perturbations around the equilibrium point.

Examples

Simple Pendulum

Assuming no damping, the differential equation governing a simple pendulum is

$$\frac{d^2\theta}{dt^2}+\frac{g}{\ell}\sin\theta=0.$$

If the maximum displacement of the pendulum is small, we can use the approximation $\sin\theta\approx\theta$ and instead consider the equation

$$\frac{d^2\theta}{dt^2}+\frac{g}{\ell}\theta=0.$$

The solution to this equation is given by:

$$\theta(t)=\theta_0\cos\left(\sqrt{\frac{g}{\ell}}t\right)$$

where θ_0 is the largest angle attained by the pendulum. The period, the time for one complete oscillation, is given by the expression

$$T_0 = 2\pi\sqrt{\frac{\ell}{g}},$$

which is a good approximation of the actual period when θ_0 is small.

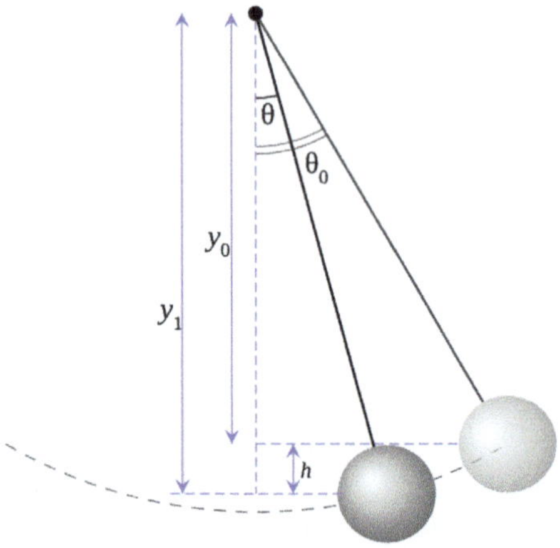

A simple pendulum exhibits approximately simple harmonic motion under the conditions of no damping

Pendulum Swinging Over Turntable

Simple harmonic motion can in some cases be considered to be the one-dimensional projection of two-dimensional circular motion. Consider a long pendulum swinging over the turntable of a record player. On the edge of the turntable there is an object. If the object is viewed from the same level as the turntable, a projection of the motion of the object seems to be moving backwards and forwards on a straight line orthogonal to the view direction, sinusoidally like the pendulum.

Spring/Mass System

When a spring is stretched or compressed by a mass, the spring develops a restoring force. Hooke's law gives the relationship of the force exerted by the spring when the spring is compressed or stretched a certain length:

$$F(t) = -kx(t)$$

where F is the force, k is the spring constant, and x is the displacement of the mass with respect to the equilibrium position. The minus sign in the equation indicates that the force exerted by the spring always acts in a direction that is opposite to the displacement (i.e. the force always acts towards the zero position), and so prevents the mass from flying off to infinity.

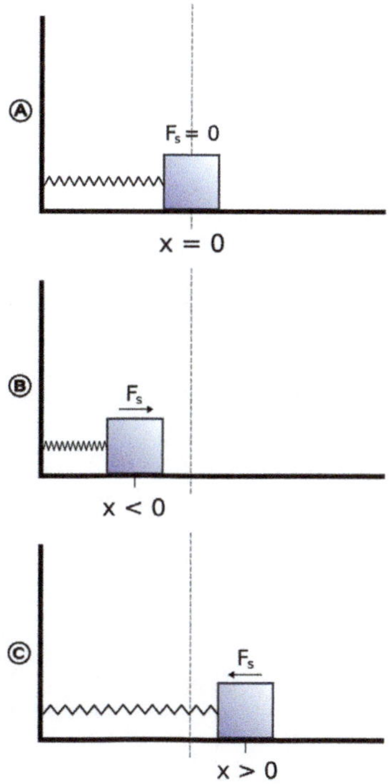

Spring–mass system in equilibrium (A), compressed (B) and stretched (C) states.

By using either force balance or an energy method, it can be readily shown that the motion of this system is given by the following differential equation:

$$F(t) = -kx(t) = m\frac{\mathrm{d}^2}{\mathrm{d}t^2}x(t) = ma.$$

...the latter being Newton's second law of motion.

If the initial displacement is A, and there is no initial velocity, the solution of this equation is given by:

$$x(t) = A\cos\left(\sqrt{\frac{k}{m}}t\right).$$

Given an ideal massless spring, m is the mass on the end of the spring. If the spring itself has mass, its effective mass must be included in m..

Energy Variation in the Spring–damping System

In terms of energy, all systems have two types of energy, potential energy and kinetic energy. When a spring is stretched or compressed, it stores elastic potential energy, which then is transferred into kinetic energy. The potential energy within a spring is determined by the equation

When the spring is stretched or compressed, kinetic energy of the mass gets converted into potential energy of the spring. By conservation of energy, assuming the datum is defined at the equilibrium position, when the spring reaches its maximum potential energy, the kinetic energy of the mass is zero. When the spring is released, it tries to return to equilibrium, and all its potential energy converts to kinetic energy of the mass.

Architectural Acoustics

Symphony Hall, Birmingham, an example of the application of architectural acoustics.

Architectural acoustics (also known as room acoustics and building acoustics) is the science and engineering of achieving a good sound within a building and is a branch of acoustical engineering. The first application of modern scientific methods to architectural acoustics was carried out by Wallace Sabine in the Fogg Museum lecture room who then applied his new found knowledge to the design of Symphony Hall, Boston.

Architectural acoustics can be about achieving good speech intelligibility in a theatre, restaurant or railway station, enhancing the quality of music in a concert hall or recording studio, or suppressing noise to make offices and homes more productive and pleasant places to work and live in. Architectural acoustic design is usually done by acoustic consultants.

Building Skin Envelope

This science analyzes noise transmission from building exterior envelope to interior and vice versa. The main noise paths are roofs, eaves, walls, windows, door and penetrations. Sufficient control ensures space functionality and is often required based on building use and local municipal codes. An example would be providing a suitable design for a home which is to be constructed close to a high volume roadway, or under the flight path of a major airport, or of the airport itself.

Inter-space Noise Control

The science of limiting and/or controlling noise transmission from one building space to another to ensure space functionality and speech privacy. The typical sound paths are ceilings, room partitions, acoustic ceiling panels (such as wood dropped ceiling panels), doors, windows, flanking, ducting and other penetrations. Technical solutions depend on the source of the noise and the path of acoustic transmission, for example noise by steps or noise by (air, water) flow vibrations. An example would be providing suitable party wall design in an apartment complex to minimize the mutual disturbance due to noise by residents in adjacent apartments.

Interior Space Acoustics

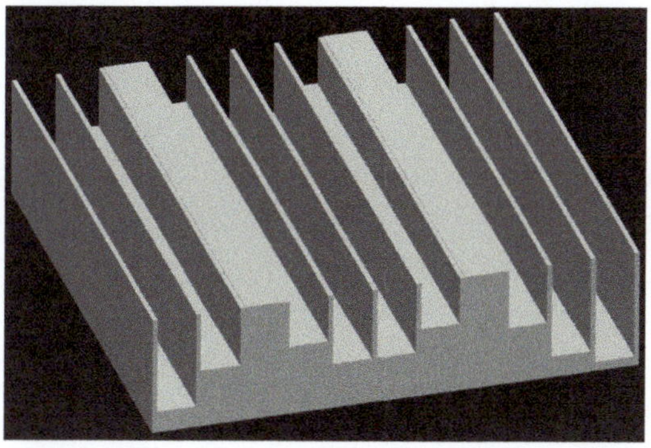

Diffusers which scatter sound are used in some rooms to improve the acoustics

This is the science of controlling a room's surfaces based on sound absorbing and reflecting properties. Excessive reverberation time, which can be calculated, can lead to poor speech intelligibility.

Ceiling of Culture Palace (Tel Aviv) concert hall is covered with perforated metal panels

Sound reflections create standing waves that produce natural resonances that can be heard as a pleasant sensation or an annoying one. Reflective surfaces can be angled and coordinated to provide good coverage of sound for a listener in a concert hall or music recital space. To illustrate this concept consider the difference between a modern large office meeting room or lecture theater and a traditional classroom with all hard surfaces.

Interior building surfaces can be constructed of many different materials and finishes. Ideal acoustical panels are those without a face or finish material that interferes with the acoustical infill or substrate. Fabric covered panels are one way to heighten acoustical absorption. Perforated metal shows also sound absorbing qualities. Finish material is used to cover over the acoustical substrate. Mineral fiber board, or Micore, is a commonly used acoustical substrate. Finish materials often consist of fabric, wood or acoustical tile. Fabric can be wrapped around substrates to create what is referred to as a "pre-fabricated panel" and often provides good noise absorption if laid onto a wall.

Prefabricated panels are limited to the size of the substrate ranging from 2 by 4 feet (0.61 m × 1.22 m) to 4 by 10 feet (1.2 m × 3.0 m). Fabric retained in a wall-mounted perimeter track system, is referred to as "on-site acoustical wall panels". This is constructed by framing the perimeter track into shape, infilling the acoustical substrate and then stretching and tucking the fabric into the perimeter frame system. On-site wall panels can be constructed to accommodate door frames, baseboard, or any other intrusion. Large panels (generally, greater than 50 square feet (4.6 m^2)) can be created on walls and ceilings with this method. Wood finishes can consist of punched or routed slots and provide a natural look to the interior space, although acoustical absorption may not be great.

There are three ways to improve workplace acoustics and solve workplace sound problems – the ABCs.

- A = Absorb (via drapes, carpets, ceiling tiles, etc.)
- B = Block (via panels, walls, floors, ceilings and layout)
- C = Cover-up (via sound masking)

Mechanical Equipment Noise

Building services noise control is the science of controlling noise produced by:

- ACMV (air conditioning and mechanical ventilation) systems in buildings, termed HVAC in North America
- Elevators
- Electrical generators positioned within or attached to a building
- Any other building service infrastructure component that emits sound.

Inadequate control may lead to elevated sound levels within the space which can be annoying and reduce speech intelligibility. Typical improvements are vibration isolation of mechanical equipment, and sound traps in ductwork. Sound masking can also be created by adjusting HVAC noise to a predetermined level.

References

- Benz Kotzen and Colin English, Environmental Noise Barriers: A Guide to Their Visual and Acoustic Design, Spon Press, United Kingdom (1999) ISBN 978-0-419-23180-6
- Randall F Barron and Barron F Barron, Industrial Noise Control and Acoustics, Marcel Dekker, New York (2002) ISBN 978-0-8247-0701-9
- Reitherman, Robert (2012). Earthquakes and Engineers: An International History. Reston, VA: ASCE Press. ISBN 9780784410714.
- Harris, C., Piersol, A., Harris Shock and Vibration Handbook, Fifth Edition, McGraw-Hill, (2002), ISBN 0-07-137081-1
- Benz Kotzen and Colin English (1999) Environmental Noise Barriers: A Guide to Their Acoustic and Visual Design, Published by Taylor & Francis, ISBN 0-419-23180-3, 165 pages
- Harris, David A. (1997). Noise Control Manual for Residential Buildings. New York: McGraw-Hill Professional. ISBN 978-0-07-026942-2.
- Serway, Raymond A.; Jewett, John W. (2003). Physics for Scientists and Engineers. Brooks/Cole. ISBN 0-534-40842-7.
- Templeton, Duncan (1993). Acoustics in the Built Environment: Advice for the Design Team. Architectural Press. ISBN 978-0750605380.
- Finz, Stacy (May 13, 2012). "High-tech system lets restaurant set noise level". San Francisco Chronicle. Archived from the original on July 16, 2013.

6

Sound and its Properties

Sound is the vibration that is audible and travels through a medium such as air or water. Elements such as pitch, duration, loudness, timbre and sound localization can analyze sound. In order to completely understand sound it is necessary to understand the properties related to it.

Sound

In physics, sound is a vibration that propagates as a typically audible mechanical wave of pressure and displacement, through a medium such as air or water. In physiology and psychology, sound is the *reception* of such waves and their *perception* by the brain. Humans can hear sound waves with frequencies between about 20 Hz and 20 kHz. Other animals have different hearing ranges.

A drum produces sound via a vibrating membrane.

Acoustics

Acoustics is the interdisciplinary science that deals with the study of mechanical waves in gases, liquids, and solids including vibration, sound, ultrasound, and infrasound. A scientist who works in the field of acoustics is an *acoustician*, while someone working in the field of acoustical engineering may be called an *acoustical engineer*. An audio engineer, on the other hand is concerned with the recording, manipulation, mixing, and reproduction of sound.

Applications of acoustics are found in almost all aspects of modern society, subdisciplines include aeroacoustics, audio signal processing, architectural acoustics, bioacoustics, electro-acoustics, environmental noise, musical acoustics, noise control, psychoacoustics, speech, ultrasound, underwater acoustics, and vibration.

Definition

Sound is defined by ANSI/ASA S1.1-2013 as "(a) Oscillation in pressure, stress, particle displacement, particle velocity, etc., propagated in a medium with internal forces (e.g., elastic or viscous), or the superposition of such propagated oscillation. (b) Auditory sensation evoked by the oscillation described in (a)." Wave is the main source of sound. Man can only listen to sound in 20 Hz-20000 Hz.

Physics of Sound

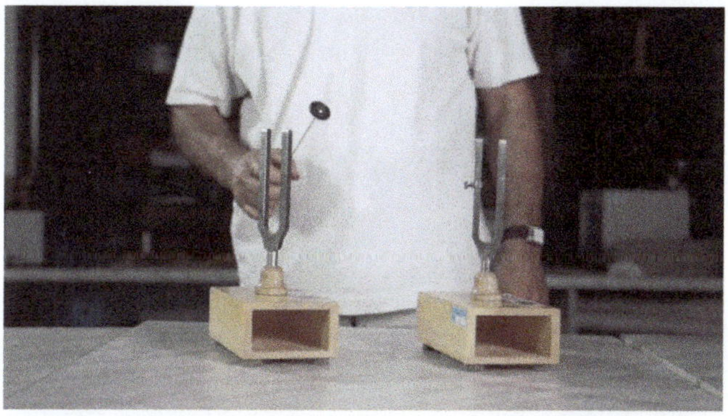

Experiment using two tuning forks oscillating at the same frequency. One of the forks is being hit with a rubberized mallet. Although the first tuning fork hasn't been hit, while the other fork is visibly excited due to the oscillation caused by the periodic change in the pressure and density of the air by hitting the other fork, creating an acoustic resonance between the forks. However, if we place a piece of metal on a prong, we see that the effect dampens, and the excitations become less and less pronounced as resonance isn't achieved as effectively. Performed by Prof. Oliver Zajkov at the Physics Institute at the Ss. Cyril and Methodius University of Skopje, Macedonia.

Sound can propagate through a medium such as air, water and solids as longitudinal waves and also as a transverse wave in solids. The sound waves are generated by a sound source, such as the vibrating diaphragm of a stereo speaker. The sound source creates vibrations in the surrounding medium. As the source continues to vibrate the medium, the vibrations propagate away from the source at the speed of sound, thus forming the sound wave. At a fixed distance from the source, the pressure, velocity, and displacement of the medium vary in time. At an instant in time, the pressure, velocity, and displacement vary in space. Note that the particles of the medium do not travel with the sound wave. This is intuitively obvious for a solid, and the same is true for liquids and gases (that is, the vibrations of particles in the gas or liquid transport the vibrations, while the *average* position of the particles over time does not change). During propagation, waves can be reflected, refracted, or attenuated by the medium.

The behavior of sound propagation is generally affected by three things:

- A complex relationship between the density and pressure of the medium. This relationship, affected by temperature, determines the speed of sound within the medium.
- Motion of the medium itself. If the medium is moving, this movement may increase or decrease the absolute speed of the sound wave depending on the direction of the movement. For example, sound moving through wind will have its speed of propagation increased by the speed of the wind if the sound and wind are moving in the same direction. If the sound and wind are moving in opposite directions, the speed of the sound wave will be decreased by the speed of the wind.
- The viscosity of the medium. Medium viscosity determines the rate at which sound is attenuated. For many media, such as air or water, attenuation due to viscosity is negligible.

When sound is moving through a medium that does not have constant physical properties, it may be refracted (either dispersed or focused).

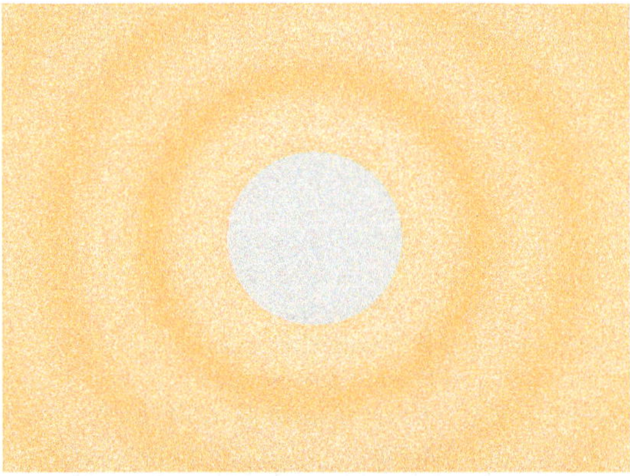

Spherical compression (longitudinal) waves

The mechanical vibrations that can be interpreted as sound are able to travel through all forms of matter: gases, liquids, solids, and plasmas. The matter that supports the sound is called the medium. Sound cannot travel through a vacuum.

Longitudinal and Transverse Waves

Sound is transmitted through gases, plasma, and liquids as longitudinal waves, also called compression waves. It requires a medium to propagate. Through solids, however, it can be transmitted as both longitudinal waves and transverse waves. Longitudinal sound waves are waves of alternating pressure deviations from the equilibrium pressure, causing local regions of compression and rarefaction, while transverse waves (in solids) are waves of alternating shear stress at right angle to the direction of propagation.

Sound waves may be "viewed" using parabolic mirrors and objects that produce sound.

The energy carried by an oscillating sound wave converts back and forth between the potential energy of the extra compression (in case of longitudinal waves) or lateral displacement strain (in case of transverse waves) of the matter, and the kinetic energy of the displacement velocity of particles of the medium.

Sound Wave Properties and Characteristics

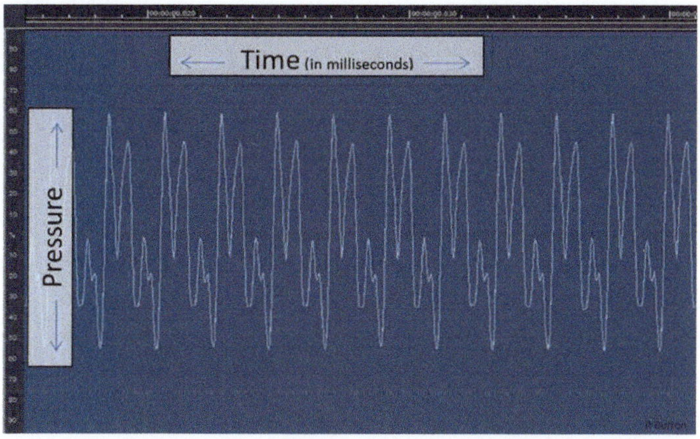

A 'pressure over time' graph of a 20 ms recording of a clarinet tone demonstrates the two fundamental elements of sound: Pressure and Time.

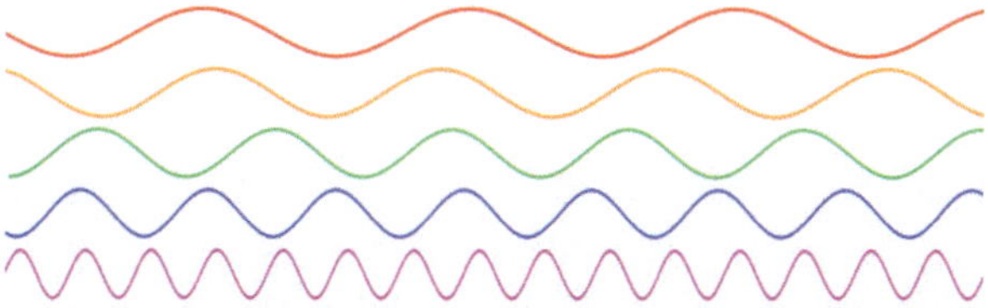

Sounds can be represented as a mixture of their component Sinusoidal waves of different frequencies. The bottom waves have higher frequencies than those above. The horizontal axis represents time.

Although there are many complexities relating to the transmission of sounds, at the point of reception (i.e. the ears), sound is readily dividable into two simple elements: pressure and time. These fundamental elements form the basis of all sound waves. They can be used to describe, in absolute terms, every sound we hear.

However, in order to understand the sound more fully, a complex wave such as this is usually separated into its component parts, which are a combination of various sound wave frequencies (and noise).

Sound waves are often simplified to a description in terms of sinusoidal plane waves, which are characterized by these generic properties:

- Frequency, or its inverse, the Wavelength
- Amplitude
- Sound pressure / Intensity
- Speed of sound
- Direction

Sound that is perceptible by humans has frequencies from about 20 Hz to 20,000 Hz. In air at standard temperature and pressure, the corresponding wavelengths of sound waves range from 17 m to 17 mm. Sometimes speed and direction are combined as a velocity vector; wave number and direction are combined as a wave vector.

Transverse waves, also known as shear waves, have the additional property, *polarization*, and are not a characteristic of sound waves.

Speed of Sound

The speed of sound depends on the medium that the waves pass through, and is a fundamental property of the material. The first significant effort towards the measure of the speed of sound was made by Newton. He believed that the speed of sound in a particular substance was equal to the square root of the pressure acting on it divided by its density:

$$c = \sqrt{\frac{p}{\rho}}$$

This was later proven wrong when found to incorrectly derive the speed. The French mathematician Laplace corrected the formula by deducing that the phenomenon of sound travelling is not isothermal, as believed by Newton, but adiabatic. He added another factor to the equation—*gamma*—and multiplied $\sqrt{\gamma}$ by $\sqrt{\frac{p}{\rho}}$, thus coming up with the equation $c = \sqrt{\gamma \cdot \frac{p}{\rho}}$. Since $K = \gamma \cdot p$, the final equation came up to be $c = \sqrt{\frac{K}{\rho}}$, which is also known as the Newton-Laplace equation. In this equation, K = elastic bulk modulus, c = velocity of sound, and ρ = density. Thus, the speed of sound is proportional to the square root of the ratio of the bulk modulus of the medium to its density.

Those physical properties and the speed of sound change with ambient conditions. For example, the speed of sound in gases depends on temperature. In 20 °C (68 °F) air at sea level, the speed of sound is approximately 343 m/s (1,230 km/h; 767 mph) using the formula "$v = (331 + 0.6\ T)$ m/s". In fresh water, also at 20 °C, the speed of sound is approximately 1,482 m/s (5,335 km/h; 3,315 mph). In steel, the speed of sound is about 5,960 m/s (21,460 km/h; 13,330 mph). The speed of sound is also slightly sensitive, being subject to a second-order anharmonic effect, to the sound amplitude, which means that there are non-linear propagation effects, such as the production of harmonics and mixed tones not present in the original sound.

Perception of Sound

A distinct use of the term *sound* from its use in physics is that in physiology and psychology, where the term refers to the subject of *perception* by the brain. The field of psychoacoustics is dedicated to such studies. Historically the word "sound" referred exclusively to an effect in the mind. Webster's 1947 dictionary defined sound as: "that which is heard; the effect which is produced by the vibration of a body affecting the ear." This meant (at least in 1947) the correct response to the question: "if a tree falls in the forest with no one to hear it fall, does it make a sound?" was "no". However, owing to contemporary usage, definitions of sound as a physical effect are prevalent in most dictionaries. Consequently, the answer to the same question is now "yes, a tree falling in the forest with no one to hear it fall does make a sound".

The physical reception of sound in any hearing organism is limited to a range of frequencies. Humans normally hear sound frequencies between approximately 20 Hz and 20,000 Hz (20 kHz), The upper limit decreases with age. Sometimes *sound* refers to only those vibrations with frequencies that are within the hearing range for humans or sometimes it relates to a particular animal. Other species have different ranges of hearing. For example, dogs can perceive vibrations higher than 20 kHz, but are deaf below 40 Hz.

As a signal perceived by one of the major senses, sound is used by many species for detecting danger, navigation, predation, and communication. Earth's atmosphere, water, and virtually any physical phenomenon, such as fire, rain, wind, surf, or earthquake, produces (and is characterized by) its unique sounds. Many species, such as frogs, birds, marine and terrestrial mammals, have also developed special organs to produce sound. In some species, these produce song and speech. Furthermore, humans have developed culture and technology (such as music, telephone and radio) that allows them to generate, record, transmit, and broadcast sound.

Elements of Sound Perception

There are six experimentally separable ways in which sound waves are analysed. They are: pitch, duration, loudness, timbre, sonic texture and spatial location.

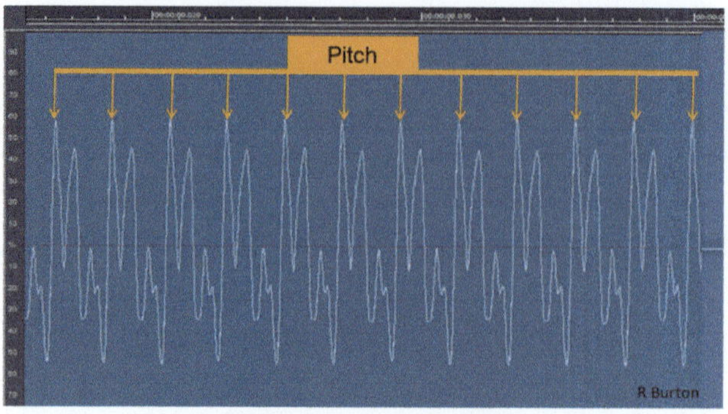

Figure. Pitch perception

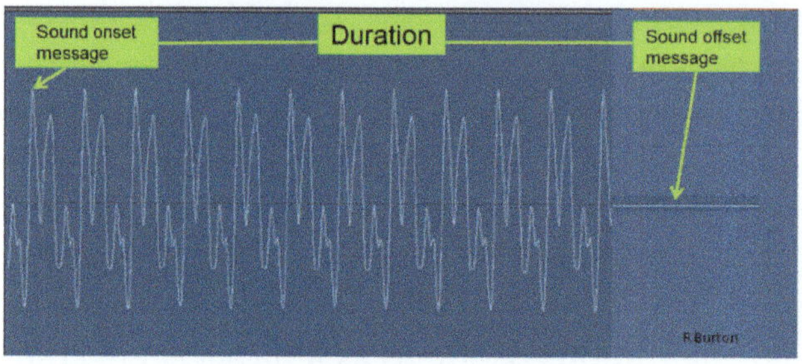

Figure. Duration perception

Pitch

Pitch is perceived as how "low" or "high" a sound is and represents the cyclic, repetitive nature of the vibrations that make up sound. For simple sounds, pitch relates to the frequency of the slowest vibration in the sound (called the fundamental harmonic). In the case of complex sounds, pitch perception can vary. Sometimes individuals identify different pitches for the same sound, based on their personal experience of particular sound patterns. Selection of a particular pitch is determined by pre-conscious examination of vibrations, including their frequencies and the balance between them. Specific attention is given to recognising potential harmonics. Every sound is placed on a pitch continuum from low to high. For example: white noise (random noise spread evenly across all frequencies) sounds higher in pitch than pink noise (random noise spread evenly across octaves) as white noise has more high frequency content. Figure 1 shows an example of pitch recognition. During the listening process, each sound is analysed for a repeating pattern and the results forwarded to the auditory cortex as a single pitch of a certain height (octave) and chroma (note name).

Duration

Duration is perceived as how "long" or "short" a sound is and relates to onset and offset signals created by nerve responses to sounds. The duration of a sound usually lasts from the time the sound is first noticed until the sound is identified as having changed or ceased. Sometimes this is not directly related to the physical duration of a sound.

For example; in a noisy environment, gapped sounds (sounds that stop and start) can sound as if they are continuous because the offset messages are missed owing to disruptions from noises in the same general bandwidth. This can be of great benefit in understanding distorted messages such as radio signals that suffer from interference, as (owing to this effect) the message is heard as if it was continuous. Figure gives an example of duration identification. When a new sound is noticed, a sound onset message is sent to the auditory cortex. When the repeating pattern is missed, a sound offset messages is sent.

Loudness

Loudness is perceived as how "loud" or "soft" a sound is and relates to the totalled number of auditory nerve stimulations over short cyclic time periods, most likely over the duration of theta wave cycles. This means that at short durations, a very short sound can sound softer than a longer sound even though they are presented at the same intensity level. Past around 200 ms this is no longer the case and the duration of the sound no longer affects the apparent loudness of the sound. Figure 3 gives an impression of how loudness information is summed over a period of about 200 ms before being sent to the auditory cortex. Louder signals create a greater 'push' on the Basilar membrane and thus stimulate more nerves,creating a stronger loudness signal. A more complex signal also creates more nerve firings and so sounds louder (for the same wave amplitude) than a simpler sound, such as a sine wave.

Timbre

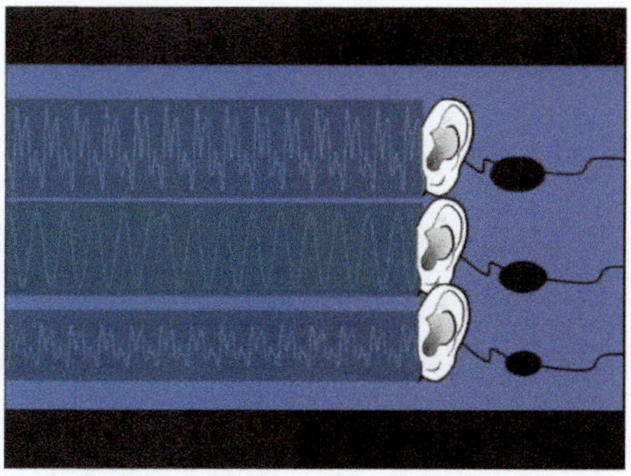

Figure. Loudness perception

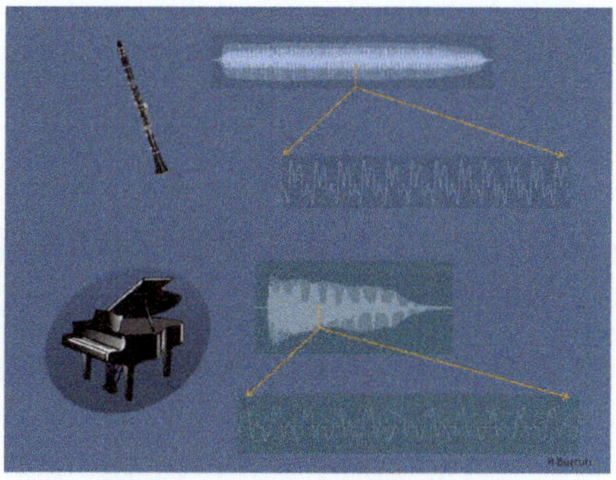

Figure. Timbre perception

Timbre is perceived as the quality of different sounds (e.g. the thud of a fallen rock, the whir of a drill, the tone of a musical instrument or the quality of a voice) and represents the pre-conscious allocation of a sonic identity to a sound (e.g. "it's an oboe!"). This identity is based on information gained from frequency transients, noisiness, unsteadiness, perceived pitch and the spread and intensity of overtones in the sound over an extended time frame. The way a sound changes over time provides most of the information for timbre identification. Even though a small section of the wave form from each instrument looks very similar, differences in changes over time between the clarinet and the piano are evident in both loudness and harmonic content. Less noticeable are the different noises heard, such as air hisses for the clarinet and hammer strikes for the piano.

Sonic Texture

Sonic texture relates to the number of sound sources and the interaction between them. The word 'texture', in this context, relates to the cognitive separation of auditory objects. In music, texture is often referred to as the difference between unison, polyphony and homophony, but it can also relate (for example) to a busy cafe; a sound which might be referred to as 'cacophony'. However texture refers to more than this. The texture of an orchestral piece is very different to the texture of a brass quartet because of the different numbers of players. The texture of a market place is very different to a school hall because of the differences in the various sound sources.

Spatial Location

Spatial location represents the cognitive placement of a sound in an environmental context; including the placement of a sound on both the horizontal and vertical plane, the distance from the sound source and the characteristics of the sonic environment. In a thick texture, it is possible to identify multiple sound sources using a combination of spatial location and timbre identification. It is the main reason why we can pick the sound of an oboe in an orchestra and the words of a single person at a cocktail party.

Noise

Noise is a term often used to refer to an unwanted sound. In science and engineering, noise is an undesirable component that obscures a wanted signal. However, in sound perception it can often be used to identify the source of a sound and is an important component of timbre perception.

Soundscape

Soundscape is the component of the acoustic environment that can be perceived by humans. The acoustic environment is the combination of all sounds (whether audible to humans or not) within a given area as modified by the environment and understood by people, in context of the surrounding environment.

Sound Pressure Level

Sound pressure is the difference, in a given medium, between average local pressure and the pressure in the sound wave. A square of this difference (i.e., a square of the deviation from the equilibrium pressure) is usually averaged over time and/or space, and a square root of this average provides a root mean square (RMS) value. For example, 1 Pa RMS sound pressure (94 dBSPL) in atmospheric air implies that the actual pressure in the sound wave oscillates between (1 atm $-\sqrt{2}$ Pa) and (1 atm $+\sqrt{2}$ Pa), that is between 101323.6 and 101326.4 Pa. As the human ear can detect sounds with a wide range of amplitudes, sound pressure is often measured as a level on a logarithmic decibel scale. The sound pressure level (SPL) or L_p is defined as

$$L_p = 10\log_{10}\left(\frac{p^2}{p_{ref}^2}\right) = 20\log_{10}\left(\frac{p}{p_{ref}}\right) \text{ dB}$$

where p is the root-mean-square sound pressure and p_{ref} is a reference sound pressure. Commonly used reference sound pressures, defined in the standard ANSI S1.1-1994, are 20 μPa in air and 1 μPa in water. Without a specified reference sound pressure, a value expressed in decibels cannot represent a sound pressure level.

Since the human ear does not have a flat spectral response, sound pressures are often frequency weighted so that the measured level matches perceived levels more closely. The International Electrotechnical Commission (IEC) has defined several weighting schemes. A-weighting attempts to match the response of the human ear to noise and A-weighted sound pressure levels are labeled dBA. C-weighting is used to measure peak levels.

Analysis of Sound

Pitch (Music)

In musical notation, the different vertical positions of notes indicate different **pitches**.

Pitch is a perceptual property of sounds that allows their ordering on a frequency-related scale, or more commonly, pitch is the quality that makes it possible to judge sounds as "higher" and "lower" in the sense associated with musical melodies. Pitch can be determined only in sounds that have a frequency that is clear and stable enough to distinguish from noise. Pitch is a major auditory attribute of musical tones, along with duration, loudness, and timbre.

Pitch may be quantified as a frequency, but pitch is not a purely objective physical property; it is a subjective psychoacoustical attribute of sound. Historically, the study of pitch and pitch perception has been a central problem in psychoacoustics, and has been instrumental in forming and testing theories of sound representation, processing, and perception in the auditory system.

Perception of Pitch

Pitch and Frequency

Pitch is an auditory sensation in which a listener assigns musical tones to relative positions on a musical scale based primarily on their perception of the frequency of vibration. Pitch is closely related to frequency, but the two are not equivalent. Frequency is an objective, scientific attribute that can be measured. Pitch is each person's *subjective perception* of a sound wave, which cannot be directly measured. However, this does not necessarily mean that most people won't agree on which notes are higher and lower.

Sound waves themselves do not have pitch, but their oscillations can be measured to obtain a frequency. It takes a sentient mind to map the internal quality of pitch. However, pitches are usually associated with, and thus quantified as frequencies in cycles per second, or hertz, by comparing sounds with pure tones, which have periodic, sinusoidal waveforms. Complex and aperiodic sound waves can often be assigned a pitch by this method.

According to the American National Standards Institute, pitch is the auditory attribute of sound according to which sounds can be ordered on a scale from low to high. Since pitch is such a close proxy for frequency, it is almost entirely determined by how quickly the sound wave is making the air vibrate and has almost nothing to do with the intensity, or amplitude, of the wave. That is, "high" pitch means very rapid oscillation, and "low" pitch corresponds to slower oscillation. Despite that, the idiom relating vertical height to sound pitch is shared by most languages. At least in English, it is just one of many deep conceptual metaphors that involve up/down. The exact etymological history of the musical sense of high and low pitch is still unclear. There is evidence that humans do actually perceive that the source of a sound is slightly higher or lower in vertical space when the sound frequency is increased or reduced.

In most cases, the pitch of complex sounds such as speech and musical notes corresponds very nearly to the repetition rate of periodic or nearly-periodic sounds, or to the reciprocal of the time interval between repeating similar events in the sound waveform.

The pitch of complex tones can be ambiguous, meaning that two or more different pitches can be perceived, depending upon the observer. When the actual fundamental frequency can be precisely determined through physical measurement, it may differ from the perceived pitch because of overtones, also known as upper partials, harmonic or otherwise. A complex tone composed of two sine waves of 1000 and 1200 Hz may sometimes be heard as up to three pitches: two spectral pitches at 1000 and 1200 Hz, derived from the physical frequencies of the pure tones, and the combination tone at 200 Hz, corresponding to the repetition rate of the waveform. In a situation like this, the percept at 200 Hz is commonly referred to as the missing fundamental, which is often the greatest common divisor of the frequencies present.

Pitch depends to a lesser degree on the sound pressure level (loudness, volume) of the tone, especially at frequencies below 1,000 Hz and above 2,000 Hz. The pitch of lower tones gets lower as sound pressure increases. For instance, a tone of 200 Hz that is very loud seems one semitone lower in pitch than if it is just barely audible. Above 2,000 Hz, the pitch gets higher as the sound gets louder.

Theories of Pitch Perception

Theories of pitch perception try to explain how the physical sound and specific physiology of the auditory system work together to yield the experience of pitch. In general, pitch perception theories can be divided into place coding and temporal coding. Place theory holds that the perception of pitch is determined by the place of maximum excitation on the basilar membrane.

A place code, taking advantage of the tonotopy in the auditory system, must be in effect for the perception of high frequencies, since neurons have an upper limit on how fast they can phase-lock their action potentials. However, a purely place-based theory cannot account for the accuracy of pitch perception in the low and middle frequency ranges.

Temporal theories offer an alternative that appeals to the temporal structure of action potentials, mostly the phase-locking and mode-locking of action potentials to frequencies in a stimulus. The precise way this temporal structure helps code for pitch at higher levels is still debated, but the processing seems to be based on an autocorrelation of action potentials in the auditory nerve. However, it has long been noted that a neural mechanism that may accomplish a delay—a necessary operation of a true autocorrelation—has not been found. At least one model shows that a temporal delay is unnecessary to produce an autocorrelation model of pitch perception, appealing to phase shifts between cochlear filters; however, earlier work has shown that certain sounds with a prominent peak in their autocorrelation function do not elicit a corresponding pitch percept, and that certain sounds without a peak in their autocorrelation function nevertheless elicit a pitch. To be a more complete model, autocorrelation must therefore apply to signals that represent the output of the cochlea, as via auditory-nerve interspike-interval histograms. Some theories of pitch perception hold that pitch has inherent octave ambiguities, and therefore is best decomposed into a pitch *chroma*, a periodic value around the octave, like the note names in western music—and a pitch *height*, which may be ambiguous, that indicates the octave the pitch is in.

Just-noticeable Difference

The *just-noticeable difference (jnd)* (the threshold at which a change is perceived) depends on the tone's frequency content. Below 500 Hz, the jnd is about 3 Hz for sine waves, and 1 Hz for complex tones; above 1000 Hz, the jnd for sine waves is about 0.6% (about 10 cents). The jnd is typically tested by playing two tones in quick succession with the listener asked if there was a difference in their pitches. The jnd becomes smaller if the two tones are played simultaneously as the listener is then able to discern beat frequencies. The total number of perceptible pitch steps in the range of human hearing is about 1,400; the total number of notes in the equal-tempered scale, from 16 to 16,000 Hz, is 120.

Aural Illusions

The relative perception of pitch can be fooled, resulting in *aural illusions*. There are several of these, such as the tritone paradox, but most notably the Shepard scale, where a continuous or discrete sequence of specially formed tones can be made to sound as if the sequence continues ascending or descending forever.

Definite and Indefinite Pitch

Not all musical instruments make notes with a clear pitch. The unpitched percussion instrument (a class of percussion instrument) does not produce particular pitches. A sound or note of definite pitch is one where a listener can possibly (or relatively easily) discern the pitch. Sounds with definite pitch have harmonic frequency spectra or close to harmonic spectra.

A sound generated on any instrument produces many modes of vibration that occur simultaneously. A listener hears numerous frequencies at once. The vibration with the lowest frequency is called the fundamental frequency; the other frequencies are overtones. Harmonics are an important class of overtones with frequencies that are integer multiples of the fundamental. Whether or not the higher frequencies are integer multiples, they are collectively called the partials, referring to the different parts that make up the total spectrum.

A sound or note of indefinite pitch is one that a listener finds impossible or relatively difficult to identify as to pitch. Sounds with indefinite pitch do not have harmonic spectra or have altered harmonic spectra—a characteristic known as inharmonicity.

It is still possible for two sounds of indefinite pitch to clearly be higher or lower than one another. For instance, a snare drum sounds higher pitched than a bass drum though both have indefinite pitch, because its sound contains higher frequencies. In other words, it is possible and often easy to roughly discern the relative pitches of two sounds of indefinite pitch, but sounds of indefinite pitch do not neatly correspond to any specific pitch. A special type of pitch often occurs in free nature when sound reaches the ear of an observer directly from the source, and also after reflecting off a sound-reflecting surface. This phenomenon is called *repetition pitch*, because the addition of a true repetition of the original sound to itself is the basic prerequisite.

Pitch Standards and Standard Pitch

A pitch standard (also Concert pitch) is the conventional pitch reference a group of musical instruments are tuned to for a performance. Concert pitch may vary from ensemble to ensemble, and has varied widely over musical history.

Standard pitch is a more widely accepted convention. The **A** above middle C is usually set at 440 Hz (often written as "A = 440 Hz" or sometimes "A440"), although other frequencies, such as 442 Hz, are also often used as variants. Another standard pitch, the so-called *Baroque pitch*, has been set in the 20th century as A = 415 Hz—approximately an equal-tempered semitone lower than A440 to facilitate transposition.

Transposing instruments have their origin in the variety of pitch standards. In modern times, they conventionally have their parts transposed into different keys from voices and other instruments (and even from each other). As a result, musicians need a way to refer to a particular pitch in an unambiguous manner when talking to each other.

For example, the most common type of clarinet or trumpet, when playing a note written in their part as C, sounds a pitch that is called B♭ on a non-transposing instrument like a violin (which indicates that at one time these wind instruments played at a standard pitch a tone lower than violin pitch). To refer to that pitch unambiguously, a musician calls it *concert B♭*, meaning, "...the pitch that someone playing a non-transposing instrument like a violin calls B♭."

Labeling Pitches

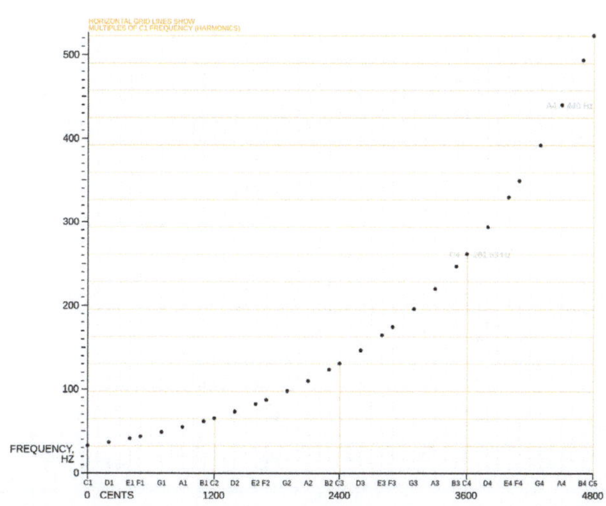

Note frequencies, four-octave C major diatonic scale, starting with C1.

Pitches are labeled using:

- Letters, as in Helmholtz pitch notation
- A combination of letters and numbers—as in scientific pitch notation, where notes are labelled upwards from C0, the 16 Hz C
- Numbers that represent the frequency in hertz (Hz), the number of cycles per second

For example, one might refer to the A above middle C as *a'*, *A4*, or *440 Hz*. In standard Western equal temperament, the notion of pitch is insensitive to "spelling": the description "G4 double sharp" refers to the same pitch as *A4*; in other temperaments, these may be distinct pitches. Human perception of musical intervals is approximately logarithmic with respect to fundamental frequency: the perceived interval between the pitches "A220" and "A440" is the same as the perceived interval between the pitches *A440* and *A880*. Motivated by this logarithmic perception, music theorists sometimes represent pitches using a numerical scale based on the logarithm of fundamental frequency. For example, one can adopt the widely used MIDI standard to map fundamental frequency, f, to a real number, p, as follows

$$p = 69 + 12 \times \log_2 \left(\frac{f}{440\ \text{Hz}} \right)$$

This creates a linear pitch space in which octaves have size 12, semitones (the distance between adjacent keys on the piano keyboard) have size 1, and A440 is assigned the number 69. Distance in this space corresponds to musical intervals as understood by musicians. An equal-tempered semitone is subdivided into 100 cents. The system is flexible enough to include "microtones" not found on standard piano keyboards. For example, the pitch halfway between C (60) and C♯ (61) can be labeled 60.5.

The following table shows frequencies in Hz for notes in various octaves, named according to the "German method" of octave nomenclature:

Note	Contra	Great	Small	One-lined	Two-lined	Three-lined
A	55.00	110.00	220.00	440.00	880.00	1760.00
A♯/B♭	58.27	116.54	233.08	466.16	932.33	1864.66
B/C♭	61.74	123.47	246.94	493.88	987.77	1975.53
B♯/C	65.41	130.81	261.63	523.25	1046.50	2093.00
C♯/D♭	69.30	138.59	277.18	554.37	1108.73	2217.46
D	73.42	146.83	293.66	587.33	1174.66	2349.32

D♯/E♭	77.78	155.56	311.13	622.25	1244.51	2489.02
E/F♭	82.41	164.81	329.63	659.26	1318.51	2637.02
E♯/F	87.31	174.61	349.23	698.46	1396.91	2793.83
F♯/G♭	92.50	185.00	369.99	739.99	1479.98	2959.96
G	98.00	196.00	392.00	783.99	1567.99	3135.96
G♯/A♭	103.83	207.65	415.30	830.61	1661.22	3322.44

Scales

The relative pitches of individual notes in a scale may be determined by one of a number of tuning systems. In the west, the twelve-note chromatic scale is the most common method of organization, with equal temperament now the most widely used method of tuning that scale. In it, the pitch ratio between any two successive notes of the scale is exactly the twelfth root of two (or about 1.05946). In well-tempered systems (as used in the time of Johann Sebastian Bach, for example), different methods of musical tuning were used. Almost all of these systems have one interval in common, the octave, where the pitch of one note is double the frequency of another. For example, if the A above middle C is 440 Hz, the A an octave above that is 880 Hz (info).

Other Musical Meanings of Pitch

In atonal, twelve tone, or musical set theory a "pitch" is a specific frequency while a pitch class is all the octaves of a frequency. In many analytic discussions of atonal and post-tonal music, pitches are named with integers because of octave and enharmonic equivalency (for example, in a serial system, C♯ and D♭ are considered the same pitch, while C4 and C5 are functionally the same, one octave apart).

Discrete pitches, rather than continuously variable pitches, are virtually universal, with exceptions including "tumbling strains" and "indeterminate-pitch chants". Gliding pitches are used in most cultures, but are related to the discrete pitches they reference or embellish.

Duration (Music)

Simple [quadr]duple drum pattern, against which duration is measured in much popular music: divides two beats into two

In music duration is an amount of time or a particular time interval: how long or short a note, phrase, section, or composition lasts. A note may last less than a second, while a symphony may last more than an hour. One of the fundamental features of rhythm, or encompassing rhythm, duration is also central to meter and musical form.

The concept of duration can be further broken down into those of beat and meter, where beat is seen as (usually, but certainly not always) a 'constant', and rhythm being longer, shorter or the same length as the beat. Pitch may even be considered a part of duration. In serial music the beginning of a note may be considered, or its duration may be (for example, is a 6 the note which begins at the sixth beat, or which lasts six beats?).

Durations, and their beginnings and endings, may be described as long, short, or taking a specific amount of time. Often duration is described according to terms borrowed from descriptions of pitch. As such, the *duration complement* is the amount of different durations used, the duration scale is an ordering (scale) of those durations from shortest to longest, the *duration range* is the difference in length between the shortest and longest, and the *duration hierarchy* is an ordering of those durations based on frequency of use.

Durational patterns are the foreground details projected against a background metric structure, which includes meter, tempo, and all rhythmic aspects which produce temporal regularity or structure. Duration patterns may be divided into rhythmic units and rhythmic gestures (Winold, 1975, chap. 3). However, they may also be described using terms borrowed from the metrical feet of poetry: iamb (weak-strong), anapest (weak-weak-strong), trochee (strong-weak), dactyl (strong-weak-weak), and amphibrach (weak-strong-weak), which may overlap to explain ambiguity.

Loudness

Loudness is the characteristic of a sound that is primarily a psycho-physiological correlate of physical strength (amplitude). More formally, it is defined as "that attribute of auditory sensation in terms of which sounds can be ordered on a scale extending from quiet to loud". The relation of physical attributes of sound to perceived loudness consists of physical, physiological and psychological components. The three components should be taken separately to permit a balanced understanding of the phenomenon.

In different industries, loudness may have different meanings, and different standards exist, each purporting to define the measurement. Some definitions such as LKFS refer to relative loudness of different segments of electronically reproduced sounds such as broadcasting and cinema. Others, such as ISO 532A (Stevens loudness, measured in sones,) ISO 532B (Zwicker loudness), DIN 45631 and ASA/ANSI S3.4, have a more general scope and are often used to characterize loudness of environmental noise. As of late 2016 there is no universally accepted method for measuring physiological loudness, as two methods are still being discussed for standardization in the next version of ISO 532B.

It is sometimes stated that loudness is a subjective measure, often confused with physical measures of sound strength such as sound pressure, sound pressure level (in decibels), sound intensity or sound power. It is often possible to separate the truly subjective components such as social considerations from the physical and physiological.

Filters such as A-weighting attempt to adjust sound measurements to correspond to loudness as perceived by the typical human, however this approach is only truly valid for loudness of single tones. A-weighting follows human sensitivity to sound and describes relative perceived loudness for at quiet to moderate speech levels, around 40 phons. However, physiological loudness perception is a much more complex process than can be captured with a single correction curve. Not only do equal-loudness contours vary with intensity, but perceived loudness of a complex sound depends on whether its spectral components are closely or widely spaced in frequency. When generating neural impulses in response to sounds of one frequency, the ear is less sensitive to nearby frequencies, which are said to be in the same critical band. Sounds containing spectral components in many critical bands are perceived as louder even if the total sound pressure remains constant.

Explanation

The perception of loudness is related to sound pressure level (SPL), frequency content and duration of a sound. The human auditory system averages the effects of SPL over a 600–1000 ms interval. A sound of constant SPL will be perceived to increase in loudness as samples of duration 20, 50, 100, 200 ms are heard, up to a duration of about 1 second at which point the perception of loudness will stabilize. For sounds of duration greater than 1 second, the moment-by-moment perception of loudness will be related to the average loudness during the preceding 600–1000 ms.

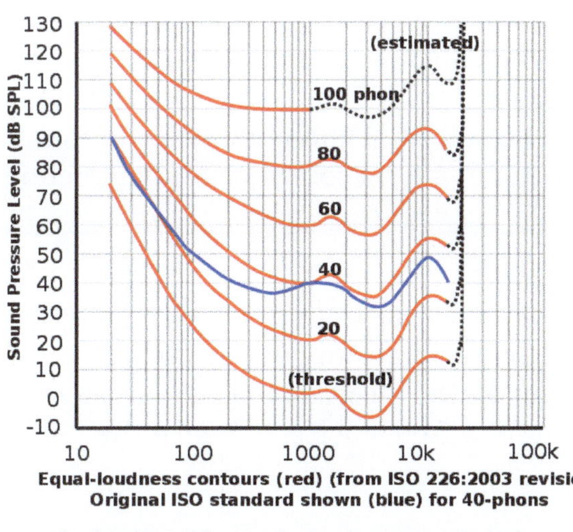

The horizontal axis shows frequency in Hz

For sounds having a duration longer than 1 second, the relationship between SPL and loudness of a single tone can be approximated by Stevens' power law in which SPL has an exponent of 0.6.[a] More precise measurements indicate that loudness increases with a higher exponent at low and high levels and with a lower exponent at moderate levels.

The sensitivity of the human ear changes as a function of frequency, as shown in the equal-loudness graph. Each line on this graph shows the SPL required for frequencies to be perceived as equally loud, and different curves pertain to different sound pressure levels. It also shows that humans with normal hearing are most sensitive to sounds around 2–4 kHz, with sensitivity declining to either side of this region. A complete model of the perception of loudness will include the integration of SPL by frequency.

Historically, loudness was measured using an "ear-balance" audiometer in which the amplitude of a sine wave was adjusted by the user to equal the perceived loudness of the sound being evaluated. Contemporary standards for measurement of loudness are based on summation of energy in critical bands as described in IEC 532, DIN 45631 and ASA/ANSI S3.4. A distinction is made between stationary loudness (sounds that remain sensibly constant) and non-stationary (sound sources that move in space or change amplitude over time.)

Hearing Loss

When sensorineural hearing loss (damage to the cochlea or in the brain) is present, the perception of loudness is altered. Sounds at low levels (often perceived by those without hearing loss as relatively quiet) are no longer audible to the hearing impaired, but sounds at high levels often are perceived as having the same loudness as they would for an unimpaired listener. This phenomenon can be explained by two theories: loudness grows more rapidly for these listeners than normal listeners with changes in level. This theory is called "loudness recruitment" and has been accepted as the classical explanation. More recently, it has been proposed that some listeners with sensorineural hearing loss may in fact exhibit a normal rate of loudness growth, but instead have an elevated loudness at their threshold. That is, the softest sound that is audible to these listeners is louder than the softest sound audible to normal listeners. This theory is called "softness imperception", a term coined by Mary Florentine.

Compensation

The "loudness" control on some consumer stereos alters the frequency response curve to correspond roughly with the equal loudness characteristic of the ear. Loudness compensation is intended to make the recorded music sound more natural when played at a lower levels by boosting low frequencies, to which the ear is less sensitive at lower sound pressure levels.

Normalization

Loudness normalization is a specific type of audio normalization that equalizes perceived level such that, for instance, commercials do not sound louder than television programs. Loudness normalization schemes exist for a number of audio applications.

Broadcast

- Commercial Advertisement Loudness Mitigation Act
- European Broadcasting Union R128

Movie and Home Theaters

- Dialnorm

Music Playback

- Sound Check in iTunes
- ReplayGain

Measurement

Historically Sone (loudness *N*) and Phon (loudness level *L*) units have been used to measure loudness.

Relative loudness monitoring in production is measured in accordance with ITU-R BS.1770 in units of LKFS.

Work began on ITU-R BS.1770 in 2001 after 0 dBFS+ level distortion in converters and lossy codecs had become evident; and the original Leq(RLB) loudness metric was proposed by Gilbert Soloudre in 2003.

Based on data from subjective listening tests, Leq(RLB) was compared against numerous other algorithms where it did remarkably well.[why?] After modification of the frequency weighting, the measurement was made multi-channel (monaural to 5.1 surround sound). CBC, Dolby and TC Electronics and numerous broadcasters contributed to the listening tests.

To make the loudness metric cross-genre friendly, a relative measurement gate was added. This work was carried out in 2008 by the EBU. The improvements were brought back into BS.1770-2. ITU subsequently updated the true-peak metric (BS.1770-3) and added provision for more audio channels, for instance 22.2 surround sound (BS.1770-4).

Timbre

In music, timbre (also known as tone color or tone quality from psychoacoustics) is the perceived sound quality of a musical note, sound, or tone that distinguishes different types of sound production, such as choir voices and musical instruments, such as string instruments, wind instruments, and percussion instruments, and which enables listeners to hear even different instruments from the same category as different (e.g. a viola and a violin).

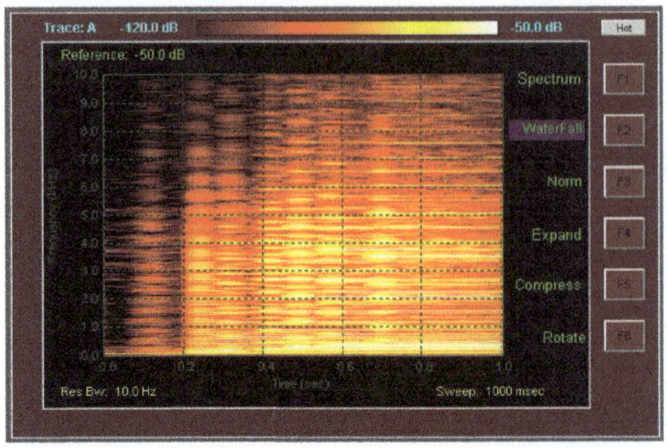

Spectrogram of the first second of an E9 suspended chord played on a Fender Stratocaster guitar with noiseless pickups. Below is the E9 suspended chord audio:

The physical characteristics of sound that determine the perception of timbre include spectrum and envelope. Singers and instrumental musicians can change the timbre of the music they are singing/playing by using different singing or playing techniques. For example, a violinist can use different bowing styles or play on different parts of the string to obtain different timbres (e.g., playing sul tasto produces a light, airy timbre, whereas playing sul ponticello produces a harsh, even an aggressive tone). On electric guitar and electric piano, performers can change the timbre using effects units and graphic equalizers.

In simple terms, timbre is what makes a particular musical sound have a different sound from another, even when they have the same pitch and loudness. For instance, it is the difference in sound between a guitar and a piano playing the same note at the same volume. Both instruments can sound equally tuned in relation to each other as they play the same note, and while playing at the same amplitude level each instrument will still sound distinctively with its own unique tone color. Experienced musicians are able to distinguish between different instruments of the same type based on their varied timbres, even if those instruments are playing notes at the same pitch and loudness.

Synonyms

Tone quality and *tone color* are synonyms for *timbre*, as well as the "*texture* attributed to a single instrument". However, the word texture can also refer to the type of music, such as multiple, interweaving melody lines versus a singable melody accompanied by subordinate chords. Hermann von Helmholtz used the German *Klangfarbe* (*tone color*), and John Tyndall proposed an English translation, *clangtint*, but both terms were disapproved of by Alexander Ellis, who also discredits *register* and *color* for their pre-existing English meanings (Erickson 1975, 7). The sound of a musical instrument may be described with words such as *bright*, *dark*, *warm*, *harsh*, and other terms. There are also colors of noise, such as pink and white. In visual representations of sound, timbre corresponds to the shape of the image (Abbado 1988, 3).

ASA Definition

The Acoustical Society of America (ASA) Acoustical Terminology definition 12.09 of timbre describes it as "that attribute of auditory sensation which enables a listener to judge that two nonidentical sounds, similarly presented and having the same loudness and pitch, are dissimilar", adding, "Timbre depends primarily upon the frequency spectrum, although it also depends upon the sound pressure and the temporal characteristics of the sound".

Attributes

Timbre has been called, "...the psychoacoustician's multidimensional waste-basket category for everything that cannot be labeled pitch or loudness." (McAdams and Bregman 1979, 34; *cf.* Dixon Ward 1965, 55 and Tobias 1970, 409).

Many commentators have attempted to decompose timbre into component attributes. For example, J. F. Schouten (1968, 42) describes the, "elusive attributes of timbre", as "determined by at least five major acoustic parameters", which Robert Erickson (1975, 5) finds, "scaled to the concerns of much contemporary music":

1. The range between tonal and noiselike character
2. The spectral envelope
3. The time envelope in terms of rise, duration, and decay (ADSR—attack, decay, sustain, release)
4. The changes both of spectral envelope (formant-glide) and fundamental frequency (micro-intonation)
5. The prefix, or onset of a sound, quite dissimilar to the ensuing lasting vibration

Erickson (1975, 6) gives a table of subjective experiences and related physical phenomena based on Schouten's five attributes:

Subjective	***Objective***
Tonal character, usually pitched	Periodic sound
Noisy, with or without some tonal character, including rustle noise	Noise, including random pulses characterized by the rustle time (the mean interval between pulses)
Coloration	Spectral envelope
Beginning/ending	Physical rise and decay time

Coloration glide or formant glide	Change of spectral envelope
Microintonation	Small change (one up and down) in frequency
Vibrato	Frequency modulation
Tremolo	Amplitude modulation
Attack	Prefix
Final sound	Suffix

Harmonics

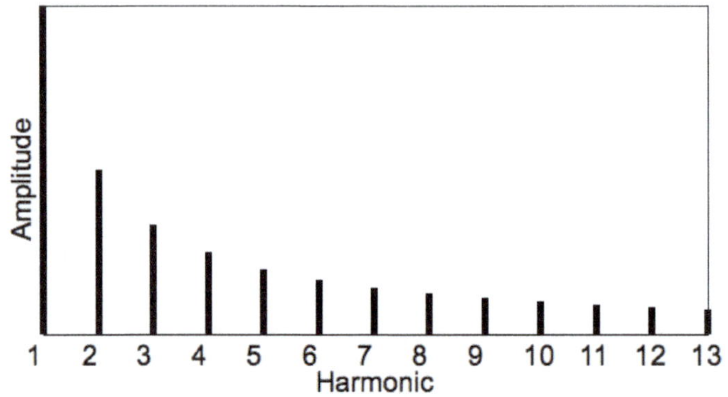

Harmonic spectra.

The richness of a sound or note a musical instrument produces is sometimes described in terms of a sum of a number of distinct frequencies. The lowest frequency is called the *fundamental frequency*, and the pitch it produces is used to name the note, but the fundamental frequency is not always the dominant frequency. The dominant frequency is the frequency that is most heard, and it is always a multiple of the fundamental frequency. For example, the dominant frequency for the transverse flute is double the fundamental frequency. Other significant frequencies are called overtones of the fundamental frequency, which may include harmonics and partials. Harmonics are whole number multiples of the fundamental frequency, such as ×2, ×3, ×4, etc. Partials are other overtones. There are also sometimes subharmonics at whole number *divisions* of the fundamental frequency. Most instruments produce harmonic sounds, but many instruments produce partials and inharmonic tones, such as cymbals and other indefinite-pitched instruments.

When the tuning note in an orchestra or concert band is played, the sound is a combination of 440 Hz, 880 Hz, 1320 Hz, 1760 Hz and so on. Each instrument in the orchestra or concert band produces a different combination of these frequencies, as well as harmonics and overtones. The sound waves of the different frequencies overlap and combine, and the balance of these amplitudes is a major factor in the characteristic sound of each instrument.

William Sethares wrote that just intonation and the western equal tempered scale are related to the harmonic spectra/timbre of many western instruments in an analogous way that the inharmonic timbre of the Thai renat (a xylophone-like instrument) is related to the seven-tone near-equal tempered pelog scale in which they are tuned. Similarly, the inharmonic spectra of Balinese metallophones combined with harmonic instruments such as the stringed rebab or the voice, are related to the five-note near-equal tempered slendro scale commonly found in Indonesian gamelan music (Sethares 1998, 6, 211, 318).

Envelope

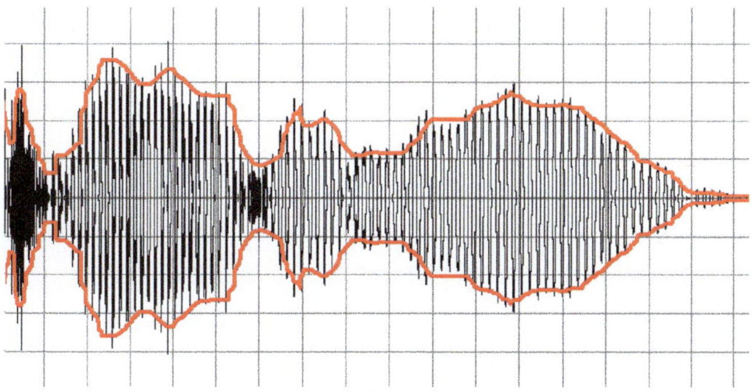

A signal and its envelope marked with red

The timbre of a sound is also greatly affected by the following aspects of its *envelope:* attack time and characteristics, decay, sustain, release (ADSR envelope) and transients. Thus these are all common controls on synthesizers. For instance, if one takes away the attack from the sound of a piano or trumpet, it becomes more difficult to identify the sound correctly, since the sound of the hammer hitting the strings or the first blast of the player's lips are highly characteristic of those instruments. The envelope is the overall amplitude structure of a sound, so called because the sound just "fits" inside its envelope: what this means should be clear from a time-domain display of almost any interesting sound, zoomed out enough that the entire waveform is visible.

In Music History

The music of Debussy, composed during the last decades of the nineteenth and the first decades of the twentieth centuries, has been credited with elevating the role of timbre in music: "To a marked degree the music of Debussy elevates timbre to an unprecedented structural status; already in *Prélude à l'après-midi d'un faune* the *color* of flute and harp functions referentially" (Samson 1977, p. 195).

In rock music from the late 1960s to the 2000s, the timbre of specific sounds is important to a song. For example, in heavy metal music, the sonic impact of the heavily-amplified, heavily-distorted power chord played on electric guitar through very loud guitar amplifiers and rows of speaker cabinets is an essential part of the style's musical identity. You could not take the heavily-amplified electric guitar part and substitute it with the same notes played on a grand piano or pipe organ.

Psychoacoustic Evidence

Often, listeners can identify an instrument, even at different pitches and loudness, in different environments, and with different players. In the case of the clarinet, acoustic analysis shows waveforms irregular enough to suggest three instruments rather than one. David Luce (1963, 16) suggests that this implies that, "Certain strong regularities in the acoustic waveform of the above instruments must exist which are invariant with respect to the above variables." However, Robert Erickson argues that there are few regularities and they do not explain our "...powers of recognition and identification." He suggests borrowing the concept of subjective constancy from studies of vision and visual perception (Erickson 1975, 11).

Psychoacoustic experiments from the 1960s onwards tried to elucidate the nature of timbre. One method involves playing pairs of sounds to listeners, then using a multi-dimensional scaling algorithm to aggregate their dissimilarity judgments into a timbre space. The most consistent outcomes from such experiments are that brightness or spectral energy distribution (Grey 1977), and the *bite*, or rate and synchronicity (Wessel 1979) and rise time (Lakatos 2000), of the attack are important factors.

Tristimulus Timbre Model

The concept of tristimulus originates in the world of color, describing the way three primary colors can be mixed together to create a given color. By analogy, the musical tristimulus measures the mixture of harmonics in a given sound, grouped into three sections. The first tristimulus measures the relative weight of the first harmonic; the second tristimulus measures the relative weight of the second, third, and fourth harmonics taken together; and the third tristimulus measures the relative weight of all the remaining harmonics (Peeters 2003; Pollard and Jansson 1982,):

$$T1 = \frac{a_1}{\sum_{h=1}^{H} a_h}$$

$$T2 = \frac{a_2 + a_3 + a_4}{\sum_{h=1}^{H} a_h}$$

$$T3 = \frac{\sum_{h=5}^{H} a_h}{\sum_{h=1}^{H} a_h}$$

Texture (Music)

Introduction to Sousa's "Washington Post March," features octave doubling (Benward & Saker 2003, p. 133) and a homorhythmic texture.

In music, texture is how the melodic, rhythmic, and harmonic materials are combined in a composition, thus determining the overall quality of the sound in a piece. Texture is often described in regard to the density, or thickness, and range, or width, between lowest and highest pitches, in relative terms as well as more specifically distinguished according to the number of voices, or parts, and the relationship between these voices. For example, a thick texture contains many 'layers' of instruments. One of these layers could be a string section, or another brass. The thickness also is changed by the amount and the richness of the instruments playing the piece. The thickness varies from light to thick. A piece's texture may be changed by the number and character of parts playing at once, the timbre of the instruments or voices playing these parts and the harmony, tempo, and rhythms used (Benward & Saker 2003). The types categorized by number and relationship of parts are analyzed and determined through the labeling of primary textural elements: primary melody (PM), secondary melody (SM), parallel supporting melody (PSM), static support (SS), harmonic support (HS), rhythmic support (RS), and harmonic and rhythmic support (HRS) (Isaac & Russell 2003, p. 136).

Common Types

In musical terms, particularly in the fields of music history and music analysis, some common terms for different types of texture are:

Type	Description	Visual
Monophonic	Monophonic texture includes a single melodic line with no accompaniment. (Benward & Saker 2003, p. 136). PSMs often double or parallel the PM they support (Benward & Saker 2003, p. 137).	 "Pop Goes the Weasel" melody (Kliewer 1975, pp. 270–301).
Biphonic	Two distinct lines, the lower sustaining a drone (constant pitch) while the other line creates a more elaborate melody above it. Pedal tones or ostinati would be an example of a SS (Benward & Saker 2003, p. 137).	 Pedal tone in Bach's Prelude no. 6 in D Minor, BWV 851, from *The Well Tempered Clavier*, Book I, mm. 1–2. All pedal tone notes are consonant except for the last three of the first measure (Benward & Saker 2003, p. 99).

Polyphonic or Counterpoint or Contrapuntal	Multiple melodic voices which are to a considerable extent independent from or in imitation with one another. Characteristic texture of the Renaissance music, also prevalent during the Baroque period (Benward & Saker 2003, pp. 1999,199,158,137, 136,129,110,90,59,35,11,9,0). Polyphonic textures may contain several PMs (Benward & Saker 2003, p. 137).	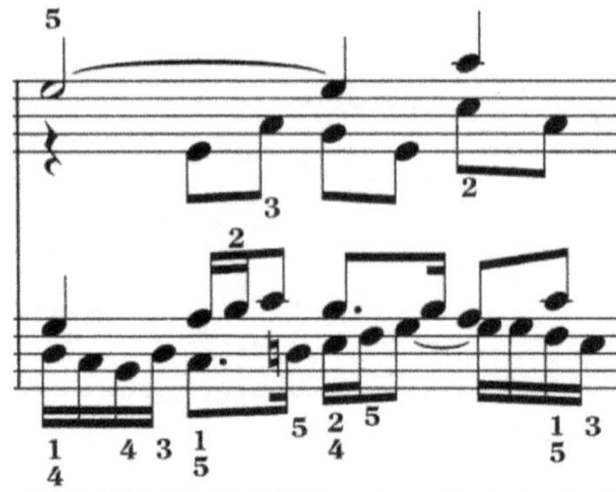 A bar from J.S. Bach's "Fugue No.17 in A flat", BWV 862, from *Das Wohltemperirte Clavier* (Part I), a famous example of contrapuntal polyphony.
Homophonic	The most common texture in Western music: melody and accompaniment. Multiple voices of which one, the melody, stands out prominently and the others form a background of harmonic accompaniment. If all the parts have much the same rhythm, the homophonic texture can also be described as homorhythmic. Characteristic texture of the Classical period and continued to predominate in Romantic music while in the 20th century, "popular music is nearly all homophonic," and, "much of jazz is also" though, "the simultaneous improvisations of some jazz musicians creates a true polyphony" (Benward & Saker 2003, p. 136). Homophonic textures usually contain only one PM (Benward & Saker 2003, p. 137). HS and RS are often combined, thus labeled HRS (Benward & Saker 2003, p. 137).	 Homophony in Tallis' "If ye love me," composed in 1549. The voices move together using the same rhythm, and the relationship between them creates chords: the excerpt begins and ends with an F major triad.
Homorhythmic	Multiple voices with similar rhythmic material in all parts. Also known as "chordal". May be considered a condition of homophony or distinguished from it.	
Heterophonic	Two or more voices simultaneously performing variations of the same melody.	

Many classical pieces feature different kinds of texture within a short space of time. An example is the Scherzo from Schubert's piano sonata in B major. The first four bars are monophonic, with both hands performing the same melody an octave apart:

Schubert Piano Sonata in B major scherzo bars 1-4 Listen

Bars 5-9 are homophonic, with all voices coinciding rhythmically:

Schubert Piano Sonata in B scherzo bars 5-10

Bars 11-20 are polyphonic. There are three parts, the top two moving in parallel (interval of a tenth). The lowest part imitates the rhythm of the upper two at the distance of three beats. The passage climaxes abruptly with a bar's silence:

Schubert Piano Sonata in B major Scherzo bars 11-20

After the silence, the polyphonic texture expands from three to four independent parts moving simultaneously in bars 21-4. The upper two parts are imitative, the lowest part consists of a repeated note (pedal point) and the remaining part weaves an independent melodic line:

Schubert Piano Sonata in B majore Scherzo bars 21-24

The final four bars revert to homophony, bringing the section to a close;

Schubert Sonata in B major Scherzo bars 25-28

Additional Types

Although in music instruction certain styles or repertoires of music are often identified with one of these descriptions this is basically added music. (for example, Gregorian chant is described as monophonic, Bach Chorales are described as homophonic and fugues as polyphonic), many composers use more than one type of texture in the same piece of music.

A simultaneity is more than one complete musical texture occurring at the same time, rather than in succession.

A more recent type of texture first used by György Ligeti is micropolyphony. Other textures include polythematic, polyrhythmic, onomatopoeic, compound, and mixed or composite textures (Corozine 2002, p. 34).

Sound localization

Sound localization refers to a listener's ability to identify the location or origin of a detected sound in direction and distance. The sound localization technique appeared with the considerable progress that has been made in acoustic simulation techniques. It allows us to localize the acoustic sources by modeling a sound field that contains one or several sources. Therefore, with such technique, people can obtain the hearing sense from any place in the sound field. It may also refer to the methods in acoustical engineering to simulate the placement of an auditory cue in a virtual 3D space.

The sound localization mechanisms of the mammalian auditory system have been extensively studied. As a new intersect discipline, sound localization involves psychological acoustics, physiological acoustics, artificial intelligence and high performance computing (HPC) system. It has been widely applied on transmitting data, information recognition and improving the immersion feelings, the fidelity and the feeling of reality of 3D environment, which can be implemented both in military and civilian areas.

The auditory system uses several cues for sound source localization, including time- and level-differences (or intensity-difference) between both ears, spectral information, timing analysis, correlation analysis, and pattern matching. These cues are also used by other animals, but there may be differences in usage, and there are also localization cues which are absent in the human auditory system, such as the effects of ear movements. Animals with the ability to localize sound have a clear evolutionary advantage.

How Sound Reaches the Brain

Sound is the perceptual result of mechanical vibrations traveling through a medium such as air or water. Through the mechanisms of compression and rarefaction, sound waves travel through the air, bounce off the pinna and concha of the exterior ear, and enter the ear canal. The sound waves vibrate the tympanic membrane (ear drum), causing the three bones of the middle ear to vibrate, which then sends the energy through the oval window and into the cochlea where it is changed into a chemical signal by hair cells in the organ of corti, which synapse onto spiral ganglion fibers that travel through the cochlear nerve into the brain.

Human Auditory System

Human ear has a very complex physical structure. As shown in fig. 1, it can be divided into three parts: outer, middle and inner.

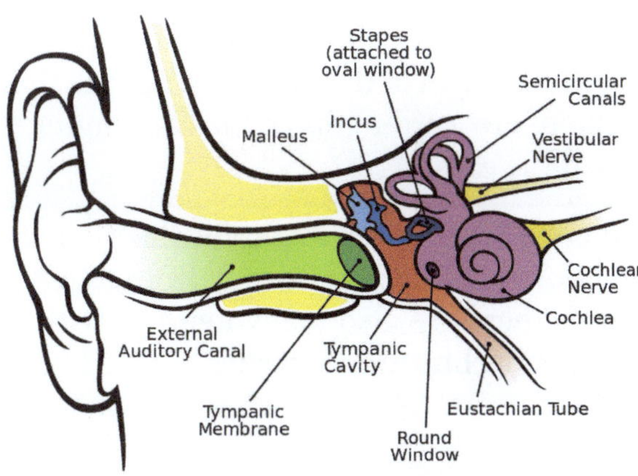

fig. 1 Human ear structure

The outer ear consists of the auricle, external auditory meatus and eardrum. For human adults, the external auditory canal is about 2.7 cm long with a diameter around 0.7 cm. When the external auditory canal is closed, the lowest resonant frequency is around 3060 Hz. Due to the resonance effect in external auditory canal, the sound will gain a 10 dB amplification. The eardrum shapes as a cone about 0.1 mm thick and 69 square mm. With normal voice level, the eardrum displacement is around 0.1 nm. In general, the out ear serves two functions: localizing the sound source and amplifying the acoustic signals.

The middle ear is a chain that consists of three pieces of ossicles: malleus, anvil and stapes. The function of middle ear is to change acoustic impedance and protect inner ear.

The main part of inner ear is cochlea. It is the receiver of human auditory systems that convert the acoustic signals to nerve stimulation.

More detailed information can be found in Ear.

Neural Interactions

In vertebrates, inter-aural time differences are known to be calculated in the superior olivary nucleus of the brainstem. According to Jeffress, this calculation relies on delay lines: neurons in the superior olive which accept innervation from each ear with different connecting axon lengths. Some cells are more directly connected to one ear than the other, thus they are specific for a particular inter-aural time difference. This theory is equivalent to the mathematical procedure of cross-correlation. However, because Jeffress' theory is unable to account for the precedence effect, in which only the first of multiple identical sounds is used to determine the sounds' location (thus avoiding confusion caused by echoes), it cannot be entirely used to explain the response. Furthermore, a number of recent physiological observations made in the midbrain and brainstem of small mammals have shed considerable doubt on the validity of Jeffress' original ideas

Neurons sensitive to inter-aural level differences (ILDs) are excited by stimulation of one ear and inhibited by stimulation of the other ear, such that the response magnitude of the cell depends on the relative strengths of the two inputs, which in turn, depends on the sound intensities at the ears.

In the auditory midbrain nucleus, the inferior colliculus (IC), many ILD sensitive neurons have response functions that decline steeply from maximum to zero spikes as a function of ILD. However, there are also many neurons with much more shallow response functions that do not decline to zero spikes.

The Cone of Confusion

Most mammals are adept at resolving the location of a sound source using interaural time differences and interaural level differences. However, no such time or level differences exist for sounds originating along the circumference of circular conical slices, where the cone's axis lies along the line between the two ears.

Consequently, sound waves originating at any point along a given circumference slant height will have ambiguous perceptual coordinates. That is to say, the listener will be incapable of determining whether the sound originated from the back, front, top, bottom or anywhere else along the circumference at the base of a cone at any given distance from the ear. Of course, the importance of these ambiguities are vanishingly small for sound sources very close to or very far away from the subject, but it is these intermediate distances that are most important in terms of fitness.

These ambiguities can be removed by tilting the head, which can introduce a shift in both the amplitude and phase of sound waves arriving at each ear. This translates the vertical orientation of the interaural axis horizontally, thereby leveraging the mechanism of localization on the horizontal plane. Moreover, even with no alternation in the angle of the interaural axis (i.e. without tilting one's head) the hearing system can capitalize on interference patterns generated by pinnae, the torso, and even the temporary re-purposing of a hand as extension of the pinna (e.g., cupping one's hand around the ear).

As with other sensory stimuli, perceptual disambiguation is also accomplished through integration of multiple sensory inputs, especially visual cues. Having localized a sound within the circumference of a circle at some perceived distance, visual cues serve to fix the location of the sound. Moreover, prior knowledge of the location of the sound generating agent will assist in resolving its current location.

Sound Localization by The Human Auditory System

Sound localization is the process of determining the location of a sound source. Objectively speaking, the major goal of sound localization is to simulate a specific sound field, including the acoustic sources, the listener, the media and environments of sound propagation. The brain utilizes subtle differences in intensity, spectral, and timing cues to allow us to localize sound sources. In this section, to more deeply understand the human auditory mechanism, we will briefly discuss about human ear localization theory.

General Introduction

Localization can be described in terms of three-dimensional position: the azimuth or horizontal angle, the elevation or vertical angle, and the distance (for static sounds) or velocity (for moving sounds).

The azimuth of a sound is signaled by the difference in arrival times between the ears, by the relative amplitude of high-frequency sounds (the shadow effect), and by the asymmetrical spectral reflections from various parts of our bodies, including torso, shoulders, and pinnae.

The distance cues are the loss of amplitude, the loss of high frequencies, and the ratio of the direct signal to the reverberated signal.

Depending on where the source is located, our head acts as a barrier to change the timbre, intensity, and spectral qualities of the sound, helping the brain orient where the sound emanated from. These minute differences between the two ears are known as interaural cues.

Lower frequencies, with longer wavelengths, diffract the sound around the head forcing the brain to focus only on the phasing cues from the source.

Helmut Haas discovered that we can discern the sound source despite additional reflections at 10 decibels louder than the original wave front, using the earliest arriving wave front. This principle is known as the Haas effect, a specific version of the precedence effect. Haas measured down to even a 1 millisecond difference in timing between the original sound and reflected sound increased the spaciousness, allowing the brain to discern the true location of the original sound. The nervous system combines all early reflections into a single perceptual whole allowing the brain to process multiple different sounds at once. The nervous system will combine reflections that are within about 35 milliseconds of each other and that have a similar intensity.

Duplex Theory

To determine the lateral input direction (left, front, right), the auditory system analyzes the following ear signal information:

Duplex Theory

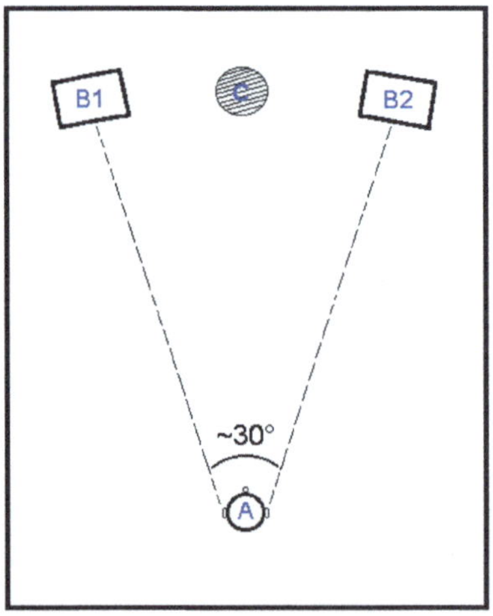

fig.2 Duplex Theory

In 1907, Lord Rayleigh utilized tuning forks to generate monophonic excitation and studied the lateral sound localization theory on a human head model without auricle. He first presented the interural clue difference based sound localization theory, which is called as Duplex Theory. Human ears are on the different sides of the head, thus they have different coordinates in space. As shown in fig. 2, since the distances between the acoustic source and ears are different, there are time difference and intensity difference between the sound signals of two ears. We call those kinds of differences as Interaural Time Difference (ITD) and Interaural Intensity Difference (IID) respectively.

ITD and IID

Interaural Time Difference (ITD) between left ear (top) and right ear (bottom).

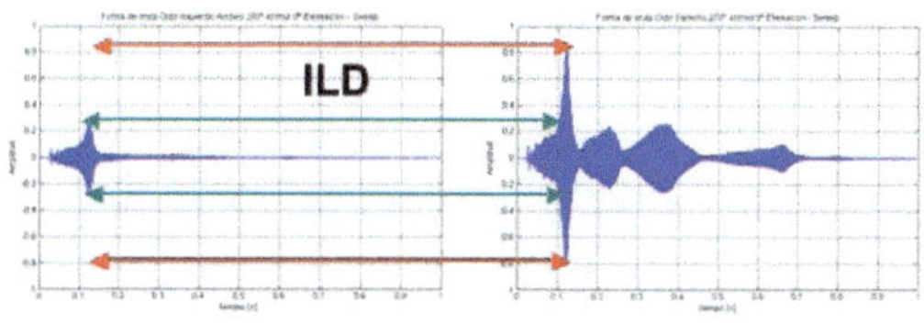

Interaural Level Difference (ILD) between left ear (left) and right ear (right).

From fig.2 we can see that no matter for source B1 or source B2, there will be a propagation delay between two ears, which will generate the ITD. Simultaneously, human head and ears may have shadowing effect on high frequency signals, which will generate IID.

- Interaural Time Difference (ITD) Sound from the right side reaches the right ear earlier than the left ear. The auditory system evaluates interaural time differences from: (a) Phase delays at low frequencies and (b) group delays at high frequencies.
- Massive experiments demonstrate that ITD relates to the signal frequency f. Suppose the angular position of the acoustic source is θ, thc hcad radius is r and the acoustic velocity is c, the function of ITD is given by:

  $ITD = \begin{cases} 300 \times \mathrm{r} \times \sin\theta / \mathrm{c}, & \text{if} f \leq 4000\text{Hz} \\ 200 \times \mathrm{r} \times \sin\theta / \mathrm{c}, & \text{if} f > 4000\text{Hz} \end{cases}$. . In above closed form, we assumed that the 0 degree is in the right ahead of the head and counter-clockwise is positive.
- Interaural Intensity Difference (IID) or Interaural Level Difference (ILD) Sound from the right side has a higher level at the right ear than at the left ear, because the head shadows the left ear. These level differences are highly frequency dependent and they increase with increasing frequency. Massive theoretical researches demonstrate that IID relates to the signal frequency f and the angular position of the acoustic source θ. The function of IID is given by: $IID = 1.0 + (f / 1000)^{0.8} \times \sin\theta$
- For frequencies below 1000 Hz, mainly ITDs are evaluated (phase delays), for frequencies above 1500 Hz mainly IIDs are evaluated. Between 1000 Hz and 1500 Hz there is a transition zone, where both mechanisms play a role.
- Localization accuracy is 1 degree for sources in front of the listener and 15 degrees for sources to the sides. Humans can discern interaural time differences of 10 microseconds or less.

Evaluation for Low Frequencies

For frequencies below 800 Hz, the dimensions of the head (ear distance 21.5 cm, corresponding to an interaural time delay of 625 μs) are smaller than the half wavelength of the sound waves. So the auditory system can determine phase delays between both ears without confusion. Interaural level differences are very low in this frequency range, especially below about 200 Hz, so a precise evaluation of the input direction is nearly impossible on the basis of level differences alone. As the frequency drops below 80 Hz it becomes difficult or impossible to use either time difference or level difference to determine a sound's lateral source, because the phase difference between the ears becomes too small for a directional evaluation.

Evaluation for High Frequencies

For frequencies above 1600 Hz the dimensions of the head are greater than the length of the sound waves. An unambiguous determination of the input direction based on interaural phase alone is not possible at these frequencies. However, the interaural level differences become larger, and these level differences are evaluated by the auditory system. Also, group delays between the ears can be evaluated, and is more pronounced at higher frequencies; that is, if there is a sound onset, the delay of this onset between the ears can be used to determine the input direction of the corresponding sound source. This mechanism becomes especially important in reverberant environments. After a sound onset there is a short time frame where the direct sound reaches the ears, but not yet the reflected sound. The auditory system uses this short time frame for evaluating the sound source direction, and keeps this detected direction as long as reflections and reverberation prevent an unambiguous direction estimation. The mechanisms described above cannot be used to differentiate between a sound source ahead of the hearer or behind the hearer; therefore additional cues have to be evaluated.

Pinna Filtering Effect Theory

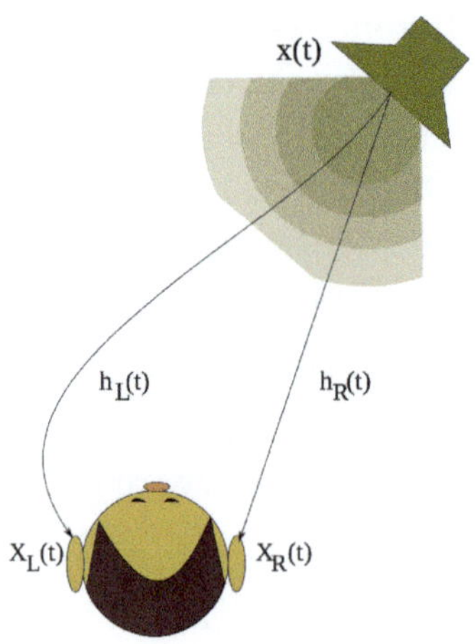

fig.4 HRTF

Motivations

Duplex theory clearly points out that ITD and IID play significant roles in sound localization but they can only deal with lateral localizing problems. For example, based on duplex theory, if two acoustic sources are symmetrically located on the right front and right back of the human head, they will generate equal ITDs and IIDs, which is called as cone model effect. However, human ears can actually distinguish this set of sources. Besides that, in natural sense of hearing, only one ear, which means no ITD or IID, can distinguish the sources with a high accuracy. Due to the disadvantages of duplex theory, researchers proposed the pinna filtering effect theory. The shape of human pinna is very special. It is concave with complex folds and asymmetrical no matter horizontally or vertically. The reflected waves and the direct waves will generate a frequency spectrum on the eardrum, which is related to the acoustic sources. Then auditory nerves localize the sources by this frequency spectrum. Therefore, a corresponding theory was proposed and called as pinna filtering effect theory.

Math Model

These spectrum clue generated by pinna filtering effect can be presented as Head-Related Transfer Functions (HRTF). The corresponding time domain expressions are called as Head-Related Impulse Response (HRIR). HRTF is also called as the transfer function from the free field to a specific point in the ear canal. We usually recognize HRTFs as LTI systems:

$$H_L = H_L(r,\theta,\varphi,\omega,\alpha) = P_L(r,\theta,\varphi,\omega,\alpha) / P_0(r,\omega)$$

$$H_R = H_R(r,\theta,\varphi,\omega,\alpha) = P_R(r,\theta,\varphi,\omega,\alpha) / P_0(r,\omega),$$

where L and R represent the left ear and right ear respectively. P_L and P_R represent the amplitude of sound pressure at entrances of left and right ear canal. P_0 is the amplitude of sound pressure at the center of the head coordinate when listener does not exist. In general, HRTFs H_L and H_R are functions of source angular position θ, elevation angle φ, distance between source and center of the head r, the angular velocity ω and the equivalent dimension of the head α.

HRTF Database

At present, the main institutes that work on measuring HRTF database includes CIPIC International Lab, MIT Media Lab, The Graduate School in Psychoacoustics at the University of Oldenburg, Neurophysiology Lab in University of Wisconsin-Madison and Ames Lab of NASA. They carefully measures the HRIRs from both humans and animals and share the results on Internet for people who want to study.

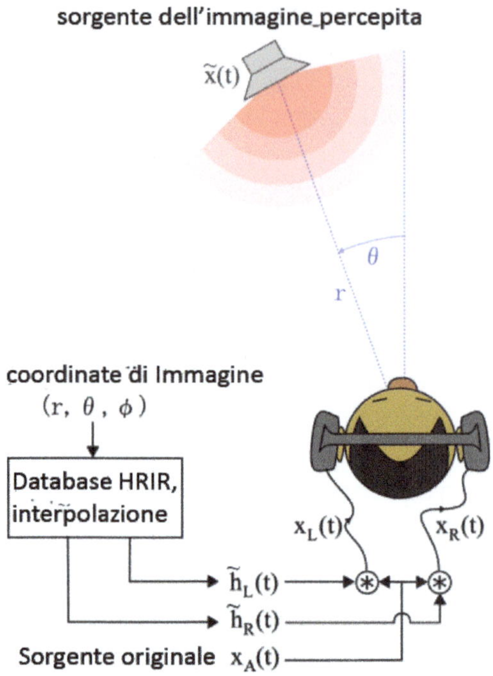

fig. 5 HRIR

Other Cues for 3D Space Localization

Monaural Cues

The human outer ear, i.e. the structures of the pinna and the external ear canal, form direction-selective filters. Depending on the sound input direction in the median plane, different filter resonances become active. These resonances implant direction-specific patterns into the frequency responses of the ears, which can be evaluated by the auditory system (directional bands) for vertical sound localization. Together with other direction-selective reflections at the head, shoulders and torso, they form the outer ear transfer functions. These patterns in the ear's frequency responses are highly individual, depending on the shape and size of the outer ear. If sound is presented through headphones, and has been recorded via another head with different-shaped outer ear surfaces, the directional patterns differ from the listener's own, and problems will appear when trying to evaluate directions in the median plane with these foreign ears. As a consequence, front–back permutations or inside-the-head-localization can appear when listening to dummy head recordings, or otherwise referred to as binaural recordings. It has been shown that human subjects can monaurally localize high frequency sound but not low frequency sound. Binaural localization, however, was possible with lower frequencies. This is likely due to the pinna being small enough to only interact with sound waves of high frequency. It seems that people can only accurately localize the elevation of sounds that are complex and include frequencies above 7,000 Hz, and a pinna must be present.

Dynamic Binaural Cues

When the head is stationary, the binaural cues for lateral sound localization (interaural time difference and interaural level difference) do not give information about the location of a sound in the median plane. Identical ITDs and ILDs can be produced by sounds at eye level or at any elevation, as long as the lateral direction is constant. However, if the head is rotated, the ITD and ILD change dynamically, and those changes are different for sounds at different elevations. For example, if an eye-level sound source is straight ahead and the head turns to the left, the sound becomes louder (and arrives sooner) at the right ear than at the left. But if the sound source is directly overhead, there will be no change in the ITD and ILD as the head turns. Intermediate elevations will produce intermediate degrees of change, and if the presentation of binaural cues to the two ears during head movement is reversed, the sound will be heard behind the listener. Hans Wallach artificially altered a sound's binaural cues during movements of the head. Although the sound was objectively placed at eye level, the dynamic changes to ITD and ILD as the head rotated were those that would be produced if the sound source had been elevated. In this situation, the sound was heard at the synthesized elevation. The fact that the sound sources objectively remained at eye level prevented monaural cues from specifying the elevation, showing that it was the dynamic change in the binaural cues during head movement that allowed the sound to be correctly localized in the vertical dimension. The head movements need not be actively produced; accurate vertical localization occurred in a similar setup when the head rotation was produced passively, by seating the blindfolded subject in a rotating chair. As long as the dynamic changes in binaural cues accompanied a perceived head rotation, the synthesized elevation was perceived.

Distance of the Sound Source

The human auditory system has only limited possibilities to determine the distance of a sound source. In the close-up-range there are some indications for distance determination, such as extreme level differences (e.g. when whispering into one ear) or specific pinna (the visible part of the ear) resonances in the close-up range.

The auditory system uses these clues to estimate the distance to a sound source:

- Direct/ Reflection ratio: In enclosed rooms, two types of sound are arriving at a listener: The direct sound arrives at the listener's ears without being reflected at a wall. Reflected sound has been reflected at least one time at a wall before arriving at the listener. The ratio between direct sound and reflected sound can give an indication about the distance of the sound source.

- Loudness: Distant sound sources have a lower loudness than close ones. This aspect can be evaluated especially for well-known sound sources.

- Sound spectrum : High frequencies are more quickly damped by the air than low frequencies. Therefore, a distant sound source sounds more muffled than a close one, because the high frequencies are attenuated. For sound with a known spectrum (e.g. speech) the distance can be estimated roughly with the help of the perceived sound.

- ITDG: The Initial Time Delay Gap describes the time difference between arrival of the direct wave and first strong reflection at the listener. Nearby sources create a relatively large ITDG, with the first reflections having a longer path to take, possibly many times longer. When the source is far away, the direct and the reflected sound waves have similar path lengths.
- Movement: Similar to the visual system there is also the phenomenon of motion parallax in acoustical perception. For a moving listener nearby sound sources are passing faster than distant sound sources.
- Level Difference: Very close sound sources cause a different level between the ears.

Signal Processing

Sound processing of the human auditory system is performed in so-called critical bands. The hearing range is segmented into 24 critical bands, each with a width of 1 Bark or 100 Mel. For a directional analysis the signals inside the critical band are analyzed together.

The auditory system can extract the sound of a desired sound source out of interfering noise. So the auditory system can concentrate on only one speaker if other speakers are also talking (the cocktail party effect). With the help of the cocktail party effect sound from interfering directions is perceived attenuated compared to the sound from the desired direction. The auditory system can increase the signal-to-noise ratio by up to 15 dB, which means that interfering sound is perceived to be attenuated to half (or less) of its actual loudness.

Localization in Enclosed Rooms

In enclosed rooms not only the direct sound from a sound source is arriving at the listener's ears, but also sound which has been reflected at the walls. The auditory system analyses only the direct sound, which is arriving first, for sound localization, but not the reflected sound, which is arriving later (law of the first wave front). So sound localization remains possible even in an echoic environment. This echo cancellation occurs in the Dorsal Nucleus of the Lateral Lemniscus (DNLL).

In order to determine the time periods, where the direct sound prevails and which can be used for directional evaluation, the auditory system analyzes loudness changes in different critical bands and also the stability of the perceived direction. If there is a strong attack of the loudness in several critical bands and if the perceived direction is stable, this attack is in all probability caused by the direct sound of a sound source, which is entering newly or which is changing its signal characteristics. This short time period is used by the auditory system for directional and loudness analysis of this sound. When reflections arrive a little bit later, they do not enhance the loudness inside the critical bands in such a strong way, but the directional cues become unstable, because there is a mix of sound of several reflection directions. As a result, no new directional analysis is triggered by the auditory system.

This first detected direction from the direct sound is taken as the found sound source direction, until other strong loudness attacks, combined with stable directional information, indicate that a new directional analysis is possible.

Specific Sound Localization Techniques with Applications

Auditory Transmission Stereo System

This kind of sound localization technique provides us the real virtual stereo system. It utilizes "smart" manikins, such as KEMAR, to glean signals or use DSP methods to simulate the transmission process from sources to ears. After amplifying, recording and transmitting, the two channels of received signals will be reproduced through earphones or speakers. This localization approach uses electroacoustic methods to obtain the spatial information of the original sound field by transferring the listener's auditory apparatus to the original sound field. The most considerable advantages of it would be that its acoustic images are lively and natural. Also, it only needs two independent transmitted signal to reproduce the acoustic image of a 3D system.

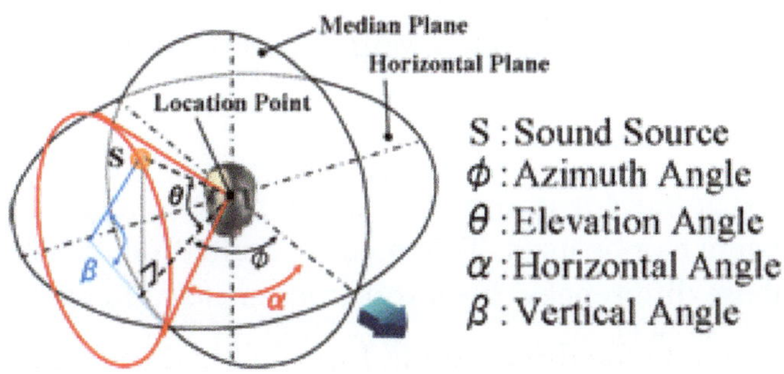

fig. 6 Sound Localization with Manikin

3D Para-virtualization Stereo System

The representatives of this this kind of system are SRS Audio Sandbox, Spatializer Audio Lab and Qsound Qxpander. They use HRTF to simulate the received acoustic signals at the ears from different directions with common binary-channel stereo reproduction. Therefore, they can simulate reflected sound waves and improve subjective sense of space and envelopment. Since they are para-virtualization stereo systems, the major goal of them is to simulate stereo sound information. Traditional stereo systems use sensors that are quite different from human ears. Although those sensors can receive the acoustic information from different directions, they do not have same the frequency response of human auditory system. Therefore, when binary-channel mode is applied, human auditory systems still cannot feel the 3D sound effect field. However, the 3D para-virtualization stereo system overcome such disadvantages. It uses HRTF principles to glean acoustic information from the original sound field then produce a lively 3D sound field through common earphones or speakers.

Multichannel Stereo Virtual Reproduction

Since the multichannel stereo systems require lots of reproduction channels, some researchers adopted the HRTF simulation technologies to reduce the number of reproduction channels. They use only two speakers to simulate multiple speakers in a multichannel system. This process is called as virtual reproduction. Essentially, such approach uses both interaural difference principle and pinna filtering effect theory. Unfortunately, this kind of approach cannot perfectly substitute the traditional multichannel stereo system, such as 5.1 surround sound system. That is because when the listening zone is relatively larger, simulation reproduction through HRTFs may cause invert acoustic images at symmetric positions.

Animals

Since most animals have two ears, many of the effects of the human auditory system can also be found in other animals. Therefore, interaural time differences (interaural phase differences) and interaural level differences play a role for the hearing of many animals. But the influences on localization of these effects are dependent on head sizes, ear distances, the ear positions and the orientation of the ears.

Lateral Information (Left, Ahead, Right)

If the ears are located at the side of the head, similar lateral localization cues as for the human auditory system can be used. This means: evaluation of interaural time differences (interaural phase differences) for lower frequencies and evaluation of interaural level differences for higher frequencies. The evaluation of interaural phase differences is useful, as long as it gives unambiguous results. This is the case, as long as ear distance is smaller than half the length (maximal one wavelength) of the sound waves. For animals with a larger head than humans the evaluation range for interaural phase differences is shifted towards lower frequencies, for animals with a smaller head, this range is shifted towards higher frequencies.

The lowest frequency which can be localized depends on the ear distance. Animals with a greater ear distance can localize lower frequencies than humans can. For animals with a smaller ear distance the lowest localizable frequency is higher than for humans.

If the ears are located at the side of the head, interaural level differences appear for higher frequencies and can be evaluated for localization tasks. For animals with ears at the top of the head, no shadowing by the head will appear and therefore there will be much less interaural level differences, which could be evaluated. Many of these animals can move their ears, and these ear movements can be used as a lateral localization cue.

Sound Localization by Odontocetes

Dolphins (and other odontocetes) rely on echolocation to aid in detecting, identifying, localizing, and capturing prey. Dolphin sonar signals are well suited for localizing multiple, small targets in a three-dimensional aquatic environment by utilizing highly directional (3 dB beamwidth of about 10 deg), broadband (3 dB bandwidth typically of about 40 kHz; peak frequencies between 40 kHz and 120 kHz), short duration clicks (about 40 μs). Dolphins can localize sounds both passively and actively (echolocation) with a resolution of about 1 deg. Cross-modal matching (between vision and echolocation) suggests dolphins perceive the spatial structure of complex objects interrogated through echolocation, a feat that likely requires spatially resolving individual object features and integration into a holistic representation of object shape. Although dolphins are sensitive to small, binaural intensity and time differences, mounting evidence suggests dolphins employ position-dependent spectral cues derived from well developed head-related transfer functions, for sound localization in both the horizontal and vertical planes. A very small temporal integration time (264 μs) allows localization of multiple targets at varying distances. Localization adaptations include pronounced asymmetry of the skull, nasal sacks, and specialized lipid structures in the forehead and jaws, as well as acoustically isolated middle and inner ears.

Sound Localization in the Median Plane (Front, Above, Back, Below)

For many mammals there are also pronounced structures in the pinna near the entry of the ear canal. As a consequence, direction-dependent resonances can appear, which could be used as an additional localization cue, similar to the localization in the median plane in the human auditory system. There are additional localization cues which are also used by animals.

Head Tilting

For sound localization in the median plane (elevation of the sound) also two detectors can be used, which are positioned at different heights. In animals, however, rough elevation information is gained simply by tilting the head, provided that the sound lasts long enough to complete the movement. This explains the innate behavior of cocking the head to one side when trying to localize a sound precisely. To get instantaneous localization in more than two dimensions from time-difference or amplitude-difference cues requires more than two detectors.

Localization with Coupled Ears (Flies)

The tiny parasitic fly Ormia ochracea has become a model organism in sound localization experiments because of its unique ear. The animal is too small for the time difference of sound arriving at the two ears to be calculated in the usual way, yet it can determine the direction of sound sources with exquisite precision. The tympanic membranes of opposite ears are directly connected mechanically, allowing resolution of sub-microsecond time differences and requiring a new neural coding strategy. Ho showed that the coupled-eardrum system in frogs can produce increased interaural vibration disparities when only small arrival time and sound level differences were available to the animal's head. Efforts to build directional microphones based on the coupled-eardrum structure are underway.

Bi-coordinate Sound Localization (Owls)

Most owls are nocturnal or crepuscular birds of prey. Because they hunt at night, they must rely on non-visual senses. Experiments by Roger Payne have shown that owls are sensitive to the sounds made by their prey, not the heat or the smell. In fact, the sound cues are both necessary and sufficient for localization of mice from a distant location where they are perched. For this to work, the owls must be able to accurately localize both the azimuth and the elevation of the sound source.

Sound Pressure

Sound pressure or acoustic pressure is the local pressure deviation from the ambient (average, or equilibrium) atmospheric pressure, caused by a sound wave. In air, sound pressure can be measured using a microphone, and in water with a hydrophone. The SI unit of sound pressure is the pascal (Pa).

Mathematical Definition

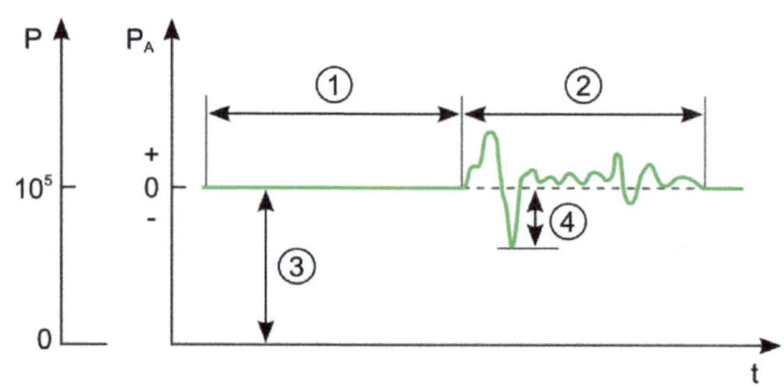

Sound pressure diagram:

1. silence; 2. audible sound; 3. atmospheric pressure; 4. sound pressure

A sound wave in a transmission medium causes a deviation (sound pressure, a *dynamic* pressure) in the local ambient pressure, a *static* pressure. Sound pressure, denoted p, is defined by

$$p_{\text{total}} = p_{\text{stat}} + p,$$

where

- p_{total} is the total pressure;
- p_{stat} is the static pressure.

Sound Measurements

Sound Intensity

In a sound wave, the complementary variable to sound pressure is the particle velocity. Together they determine the sound intensity of the wave. *Sound intensity*, denoted $\mathbf{I}$ and measured in W·m^{-2} in SI units, is defined by

$$\mathbf{I} = p\mathbf{v},$$

where

- p is the sound pressure;
- $\mathbf{v}$ is the particle velocity.

Acoustic Impedance

Acoustic impedance, denoted Z and measured in Pa·m^{-3}·s in SI units, is defined by

$$Z(s) = \frac{\hat{p}(s)}{\hat{Q}(s)},$$

where

- $\hat{p}(s)$ is the Laplace transform of sound pressure;
- $\hat{Q}(s)$ is the Laplace transform of sound volume flow rate.

Specific acoustic impedance, denoted z and measured in Pa·m^{-1}·s in SI units, is defined by

$$z(s) = \frac{\hat{p}(s)}{\hat{v}(s)},$$

where

- $\hat{p}(s)$ is the Laplace transform of sound pressure;
- $\hat{v}(s)$ is the Laplace transform of particle velocity.

Particle Displacement

The *particle displacement* of a *progressive sine wave* is given by

$$\delta(\mathbf{r},t) = \delta_{\mathrm{m}} \cos(\mathbf{k}\cdot\mathbf{r} - \omega t + \varphi_{\delta,0}),$$

where

- δ_{m} is the amplitude of the particle displacement;
- $\varphi_{\delta,0}$ is the phase shift of the particle displacement;
- k is the angular wavevector;
- ω is the angular frequency.

It follows that the particle velocity and the sound pressure along the direction of propagation of the sound wave x are given by

$$v(\mathbf{r},t) = \frac{\partial \delta}{\partial t}(\mathbf{r},t) = \omega\delta_{\mathrm{m}} \cos\left(\mathbf{k}\cdot\mathbf{r} - \omega t + \varphi_{\delta,0} + \frac{\pi}{2}\right) = v_{\mathrm{m}} \cos(\mathbf{k}\cdot\mathbf{r} - \omega t + \varphi_{v,0}),$$

$$p(\mathbf{r},t) = -\rho c^2 \frac{\partial \delta}{\partial x}(\mathbf{r},t) = \rho c^2 k_x \delta_{\mathrm{m}} \cos\left(\mathbf{k}\cdot\mathbf{r} - \omega t + \varphi_{\delta,0} + \frac{\pi}{2}\right) = p_{\mathrm{m}} \cos(\mathbf{k}\cdot\mathbf{r} - \omega t + \varphi_{p,0}),$$

where

- v_{m} is the amplitude of the particle velocity;
- $\varphi_{v,0}$ is the phase shift of the particle velocity;
- p_{m} is the amplitude of the acoustic pressure;
- $\varphi_{p,0}$ is the phase shift of the acoustic pressure.

Taking the Laplace transforms of v and p with respect to time yields

$$\hat{v}(\mathbf{r},s) = v_{\mathrm{m}} \frac{s\cos\varphi_{v,0} - \omega\sin\varphi_{v,0}}{s^2 + \omega^2},$$

$$\hat{p}(\mathbf{r},s) = p_{\mathrm{m}} \frac{s\cos\varphi_{p,0} - \omega\sin\varphi_{p,0}}{s^2 + \omega^2}.$$

Since $\varphi_{v,0}$ $\varphi_{p,0}$, the amplitude of the specific acoustic impedance is given by

$$z_{\mathrm{m}}(\mathbf{r},s) = |z(\mathbf{r},s)| = \left|\frac{\hat{p}(\mathbf{r},s)}{\hat{v}(\mathbf{r},s)}\right| = \frac{p_{\mathrm{m}}}{v_{\mathrm{m}}} = \frac{\rho c^2 k_x}{\omega}.$$

Consequently, the amplitude of the particle displacement is related to that of the acoustic velocity and the sound pressure by

$$\delta_{\mathrm{m}} = \frac{v_{\mathrm{m}}}{\omega},$$

$$\delta_{\mathrm{m}} = \frac{p_{\mathrm{m}}}{\omega z_{\mathrm{m}}(\mathbf{r},s)}.$$

Inverse-proportional Law

When measuring the sound pressure created by an object, it is important to measure the distance from the object as well, since the sound pressure of a *spherical* sound wave decreases as $1/r$ from the centre of the sphere (and not as $1/r^2$, like the sound intensity):

$$p(r) \propto \frac{1}{r}.$$

This relationship is an *inverse-proportional law.*

If the sound pressure p_1 is measured at a distance r_1 from the centre of the sphere, the sound pressure p_2 at another position r_2 can be calculated:

$$p_2 = \frac{r_1}{r_2} p_1.$$

The inverse-proportional law for sound pressure comes from the inverse-square law for sound intensity:

$$I(r) \propto \frac{1}{r^2}.$$

Indeed,

$$I(r) = p(r)v(r) = p(r)[p * z^{-1}](r) \propto p^2(r),$$

where

- $*$ is the convolution operator;
- z^{-1} is the convolution inverse of the specific acoustic impedance,

hence the inverse-proportional law:

$$p(r) \propto \frac{1}{r}.$$

The sound pressure may vary in direction from the centre of the sphere as well, so measurements at different angles may be necessary, depending on the situation. An obvious example of a sound source whose spherical sound wave varies in level in different directions is a bullhorn.

Sound Pressure Level

Sound pressure level (SPL) or acoustic pressure level is a logarithmic measure of the effective pressure of a sound relative to a reference value. Sound pressure level, denoted L_p and measured in dB, is defined by

$$L_p = \ln\left(\frac{p}{p_0}\right)\text{Np} = 2\log_{10}\left(\frac{p}{p_0}\right)\text{B} = 20\log_{10}\left(\frac{p}{p_0}\right)\text{dB},$$

where

- p is the root mean square sound pressure;
- p_0 is the *reference sound pressure*;
- 1 Np is the neper;
- 1 B = (1/2 ln 10) Np is the bel;
- 1 dB = (1/20 ln 10) Np is the decibel.

The commonly used reference sound pressure in air is

$$p_0 = 20\,\mu\text{Pa},$$

which is often considered as the threshold of human hearing (roughly the sound of a mosquito flying 3 m away). The proper notations for sound pressure level using this reference are $L_{p/(20\,\mu\text{Pa})}$ or L_p (re 20 μPa), but the suffix notations dB SPL, dB(SPL), dB-SPL, or dB_{SPL} are very common, even if they are not accepted by the SI.

Most sound level measurements will be made relative to this reference, meaning 1 Pa will equal an SPL of 94 dB. In other media, such as underwater, a reference level of 1 μPa is used. These references are defined in ANSI S1.1-1994.

Examples

The lower limit of audibility is defined as SPL of 0 dB, but the upper limit is not as clearly defined. While 1 atm (194 dB Peak or 191 dB SPL) is the largest pressure variation an undistorted sound wave can have in Earth's atmosphere, larger sound waves can be present in other atmospheres or other media such as under water, or through the Earth.

Ears detect changes in sound pressure. Human hearing does not have a flat spectral sensitivity (frequency response) relative to frequency versus amplitude. Humans do not perceive low- and high-frequency sounds as well as they perceive sounds between 3,000 and 4,000 Hz, as shown in the equal-loudness contour. Because the frequency response of human hearing changes with amplitude, three weightings have been established for measuring sound pressure: A, B and C. A-weighting applies to sound pressures levels up to 55 dB, B-weighting applies to sound pressures levels between 55 dB and 85 dB, and C-weighting is for measuring sound pressure levels above 85 dB.

In order to distinguish the different sound measures a suffix is used: A-weighted sound pressure level is written either as dB_A or L_A. B-weighted sound pressure level is written either as dB_B or L_B, and C-weighted sound pressure level is written either as dB_C or L_C. Unweighted sound pressure level is called "linear sound pressure level" and is often written as dB_L or just L. Some sound measuring instruments use the letter "Z" as an indication of linear SPL.

Distance

The distance of the measuring microphone from a sound source is often omitted when SPL measurements are quoted, making the data useless. In the case of ambient environmental measurements of "background" noise, distance need not be quoted as no single source is present, but when measuring the noise level of a specific piece of equipment the distance should always be stated. A distance of one metre (1 m) from the source is a frequently used standard distance. Because of the effects of reflected noise within a closed room, the use of an anechoic chamber allows for sound to be comparable to measurements made in a free field environment.

According to the inverse proportional law, when sound level L_{p1} is measured at a distance r_1, the sound level L_{p2} at the distance r_2 is

$$L_{p_2} = L_{p_1} + 20\log_{10}\left(\frac{r_1}{r_2}\right)\text{dB}.$$

Multiple Sources

The formula for the sum of the sound pressure levels of n incoherent radiating sources is

$$L_{\acute{O}} = 10\log_{10}\left(\frac{p_1^{\ 2} + p_2^{\ 2} + \ldots + p_n^{\ 2}}{p_0^{\ 2}}\right)\text{dB} = 10\log_{10}\left[\left(\frac{p_1}{p_0}\right)^2 + \left(\frac{p_2}{p_0}\right)^2 + \ldots + \left(\frac{p_n}{p_0}\right)^2\right]\text{dB}.$$

Inserting the formulas

$$\left(\frac{p_i}{p_0}\right)^2 = 10^{\frac{L_i}{10\text{dB}}}, \quad i = 1,2,\ldots,n,$$

in the formula for the sum of the sound pressure levels yields

$$L_{\Sigma} = 10\log_{10}\left(10^{\frac{L_1}{10\text{dB}}} + 10^{\frac{L_2}{10\text{dB}}} + \ldots + 10^{\frac{L_n}{10\text{dB}}}\right)\text{dB}.$$

Examples of Sound Pressure

Examples of sound pressure in air at standard atmospheric pressure			
Source of sound	**Distance**	**Sound pressure* (Pa)**	**Sound pressure level (dB_{SPL})**
Shockwave (distorted sound waves > 1 atm; waveform valleys are clipped at zero pressure)		>101,325	>191
Theoretical limit for undistorted sound at 1 atmosphere environmental pressure		101,325	191

Stun grenade	Ambient	1,600–8,000	158–172
Simple open-ended thermoacoustic device		12,619	176
.30-06 rifle being fired	1 m to shooter's side	7,265	171
LRAD 1000Xi Long Range Acoustic Device	1 m	893	153
Jet engine	1 m	632	150
Threshold of pain	At ear	63.2-200	130-140
Loudest human voice	1 inch	110	135
Trumpet	0.5 m	63.2	130
Vuvuzela horn	1 m	20	120
Risk of instantaneous noise-induced hearing loss	At ear	20	120
Jet engine	100 m	6.32–200	110–140
Non-electric chainsaw	1 m	6.32	110
Jack hammer	1 m	2	100
Traffic on a busy roadway	10 m	0.2–0.632	80–90
Hearing damage (over long-term exposure, need not be continuous)	At ear	0.356	85
Passenger car	10 m	$(2–20)\times10^{-2}$	60–80
EPA-identified maximum to protect against hearing loss and other disruptive effects from noise, such as sleep disturbance, stress, learning detriment, etc.	Ambient	6.32×10^{-2}	70
TV (set at home level)	1 m	2×10^{-2}	60
Normal conversation	1 m	$(2–20)\times10^{-3}$	40–60
Very calm room	Ambient	$(2–6.32)\times10^{-4}$	20–30
Light leaf rustling, calm breathing	Ambient	6.32×10^{-5}	10
Auditory threshold at 1 kHz	At ear	2×10^{-5}	0
Anechoic chamber, Orfeld Labs (A-weighted)	Ambient	6.8×10^{-6}	−9.4
Anechoic chamber, University of Salford (A-weighted)	Ambient	4.8×10^{-6}	−12.4
Anechoic chamber, Microsoft (A-weighted)	Ambient	1.9×10^{-6}	−20.35

*All values listed are the effective sound pressure unless otherwise stated.

Sound Power

Sound power or acoustic power is the rate at which sound energy is emitted, reflected, transmitted or received, per unit time. The SI unit of sound power is the watt (W). It is the power of the sound force on a surface of the medium of propagation of the sound wave. For a sound source, unlike sound pressure, sound power is neither room-dependent nor distance-dependent. Sound pressure is a measurement at a point in space near the source, while the sound power of a source is the total power emitted by that source in all directions. Sound power passing through an area is sometimes called sound flux or acoustic flux through that area.

Mathematical Definition

Sound power, denoted P, is defined by

$$P = \mathbf{f} \cdot \mathbf{v} = Ap\mathbf{u} \cdot \mathbf{v} = Apv$$

where

- $\mathbf{f}$ is the sound force of unit vector $\mathbf{u}$;
- $\mathbf{v}$ is the particle velocity of projection v along $\mathbf{u}$;
- A is the area;
- p is the sound pressure.

In a medium, the sound power is given by

$$P = \frac{Ap^2}{\rho c} \cos\theta,$$

where

- A is the area of the surface;
- ρ is the mass density;
- c is the sound velocity;
- θ is the angle between the direction of propagation of the sound and the normal to the surface.

For example, a sound at SPL = 85 dB or p = 0.356 Pa in air (ρ = 1.2 kg·m^{-3} and c = 343 m·s^{-1}) through a surface of area A = 1 m^2 normal to the direction of propagation (θ = 0 °) has a sound energy flux P = 0.3 mW.

This is the parameter one would be interested in when converting noise back into usable energy, along with any losses in the capturing device.

Table of Selected Sound Sources

Maximum sound power level (L_{WA}) related to a portable air compressor.

Here is a table of some examples, from an on-line source. The top three or four lines seem questionable. A 150-dB source should be audible 100 times as far away as a 110-dB source, and a 200-dB source 10 000 times further than a 120-dB source. This would imply that a turbofan aircraft at take-off would be audible 100 times further than a rock concert, and a Saturn V 10 000 times further than heavy thunder. However this simple analysis does not account for the effect of atmospheric absorption effects and how they change with frequency.

Situation and sound source	Sound power (W)	Sound power level (dB ref 10^{-12} W)
Saturn V rocket	**100,000,000**	200
Project Artemis Sonar	**1,000,000**	
Turbojet engine	**100,000**	170
Turbofan aircraft at take-off	**1,000**	150
Turboprop aircraft at take-off	**100**	140
Machine gun Large pipe organ	**10**	130
Symphony orchestra Heavy thunder Sonic boom	**1**	120
Rock concert Chain saw Accelerating motorcycle	**0.1**	110
Lawn mower Car at highway speed Subway steel wheels	**0.01**	100

Large diesel vehicle	**0.001**	90
a loud Alarm clock	**0.0001**	80
a relatively quiet Vacuum cleaner	**10^{-5}**	70
Hair dryer	**10^{-6}**	60
Radio or TV	**10^{-7}**	50
Refrigerator low voice	**10^{-8}**	40
Quiet conversation	**10^{-9}**	30
Whisper of one person Wristwatch ticking	**10^{-10}**	20
Human breath of one person	**10^{-11}**	10
Reference value	**10^{-12}**	0

Relationships with Other Quantities

Sound power is related to sound intensity:

$$P = AI,$$

where

- A is the area;
- I is the sound intensity.

Sound power is related sound energy density:

$$P = Acw,$$

where

- c is the speed of sound;
- w is the sound energy density.

Sound Power Level

Sound power level (SWL) or acoustic power level is a logarithmic measure of the power of a sound relative to a reference value. Sound power level, denoted L_W and measured in dB, is defined by

$$L_W = \frac{1}{2}\ln\left(\frac{P}{P_0}\right)\text{Np} = \log_{10}\left(\frac{P}{P_0}\right)\text{B} = 10\log_{10}\left(\frac{P}{P_0}\right)\text{dB},$$

where

- P is the sound power;

- P_0 is the *reference sound power*;
- 1 Np = 1 is the neper;
- 1 B = 1/2 ln 10 is the bel;
- 1 dB = 1/20 ln 10 is the decibel.

The commonly used reference sound power in air is

$$P_0 = 1\,\text{pW}.$$

The proper notations for sound power level using this reference are $L_{W/(1\,\text{pW})}$ or L_W (re 1 pW), but the suffix notations dB SWL, dB(SWL), dBSWL, or dB_{SWL} are very common, even if they are not accepted by the SI.

The reference sound power P_0 is defined as the sound power with the reference sound intensity I_0 = 1 pW/m² passing through a surface of area A_0 = 1 m²:

$$P_0 = A_0 I_0,$$

hence the reference value P_0 = 1 pW.

Relationship with Sound Pressure Level

The generic calculation of sound power from sound pressure is as follows:

$$L_W = L_p + 10\log_{10}\left(\frac{A_S}{A_0}\right)\text{dB},$$

where: A_S defines the area of a surface that wholly encompasses the source. This surface may be any shape, but it must fully enclose the source.

In the case of a sound source located in free field positioned over a reflecting plane (i.e. the ground), in air at ambient temperature, the sound power level at distance r from the sound source is approximately related to sound pressure level (SPL) by

$$L_W = L_p + 10\log_{10}\left(\frac{2\pi r^2}{A_0}\right)\text{dB},$$

where

- L_p is the sound pressure level;
- A_0 = 1 m²;
- $2\pi r^2$, defines the surface area of a hemisphere; and
- r must be sufficient that the hemisphere fully encloses the source.

Derivation of this equation:

$$L_W = \frac{1}{2}\ln\left(\frac{P}{P_0}\right)$$
$$= \frac{1}{2}\ln\left(\frac{AI}{A_0 I_0}\right)$$
$$= \frac{1}{2}\ln\left(\frac{I}{I_0}\right) + \frac{1}{2}\ln\left(\frac{A}{A_0}\right).$$

For a *progressive* spherical wave,

$$z_0 = \frac{p}{v},$$
$$A = 4\pi r^2, \text{ (the surface area of sphere)}$$

where z_0 is the characteristic specific acoustic impedance.

Consequently,

$$I = pv = \frac{p^2}{z_0},$$

and since by definition $I_0 = p_0^2/z_0$, where $p_0 = 20$ µPa is the reference sound pressure,

$$L_W = \frac{1}{2}\ln\left(\frac{p^2}{p_0^2}\right) + \frac{1}{2}\ln\left(\frac{4\pi r^2}{A_0}\right) =$$
$$\ln\left(\frac{p}{p_0}\right) + \frac{1}{2}\ln\left(\frac{4\pi r^2}{A_0}\right)$$
$$= L_p + 10\log_{10}\left(\frac{4\pi r^2}{A_0}\right)\text{dB}.$$

The sound power estimated practically does not depend on distance. The sound pressure used in the calculation may be affected by distance due to viscous effects in the propagation of sound unless this is accounted for.

Speed of Sound

The speed of sound is the distance travelled per unit time by a sound wave as it propagates through an elastic medium. In dry air at 20 °C (68 °F), the speed of sound is 343.2 metres per second (1,126 ft/s; 1,236 km/h; 768 mph; 667 kn), or a kilometre in 2.914 s or a mile in 4.689 s.

The speed of sound in an ideal gas depends only on its temperature and composition. The speed has a weak dependence on frequency and pressure in ordinary air, deviating slightly from ideal behavior.

In common everyday speech, *speed of sound* refers to the speed of sound waves in air. However, the speed of sound varies from substance to substance: sound travels most slowly in gases; it travels faster in liquids; and faster still in solids. For example (as noted above), sound travels at 343.2 m/s in air; it travels at 1,484 m/s in water (4.3 times as fast as in air); and at 5,120 m/s in iron. In an exceptionally stiff material such as diamond, sound travels at 12,000 m/s; which is around the maximum speed that sound will travel under normal conditions.

Sound waves in solids are composed of compression waves (just as in gases and liquids), but there is also a different type of sound wave called a shear wave, which occurs only in solids. These different types of waves in solids usually travel at different speeds, as exhibited in seismology. The speed of a compression sound wave in solids is determined by the medium's compressibility, shear modulus and density. The speed of shear waves is determined only by the solid material's shear modulus and density.

In fluid dynamics, the speed of sound in a fluid medium (gas or liquid) is used as a relative measure for the speed of an object moving through the medium. The ratio of the speed of an object to the speed of sound in the fluid is called the object's Mach number. Objects moving at speeds greater than *Mach1* are said to be traveling at supersonic speeds.

History

Sir Isaac Newton computed the speed of sound in air as 979 feet per second (298 m/s), which is too low by about 15%, but had neglected the effect of fluctuating temperature; that was later rectified by Laplace.

During the 17th century, there were several attempts to measure the speed of sound accurately, including attempts by Marin Mersenne in 1630 (1,380 Parisian feet per second), Pierre Gassendi in 1635 (1,473 Parisian feet per second) and Robert Boyle (1,125 Parisian feet per second).

In 1709, the Reverend William Derham, Rector of Upminster, published a more accurate measure of the speed of sound, at 1,072 Parisian feet per second. Derham used a telescope from the tower of the church of St Laurence, Upminster to observe the flash of a distant shotgun being fired, and then measured the time until he heard the gunshot with a half second pendulum. Measurements were made of gunshots from a number of local landmarks, including North Ockendon church. The distance was known by triangulation, and thus the speed that the sound had travelled was calculated.

Basic Concept

The transmission of sound can be illustrated by using a model consisting of an array of balls interconnected by springs. For real material the balls represent molecules and the springs represent the bonds between them. Sound passes through the model by compressing and expanding the springs, transmitting energy to neighbouring balls, which transmit energy to *their* springs, and so on. The speed of sound through the model depends on the stiffness of the springs, and the mass of the balls. As long as the spacing of the balls remains constant, stiffer springs transmit energy more quickly, and more massive balls transmit energy more slowly. Effects like dispersion and reflection can also be understood using this model.

In a real material, the stiffness of the springs is called the elastic modulus, and the mass corresponds to the density. All other things being equal (ceteris paribus), sound will travel more slowly in spongy materials, and faster in stiffer ones. For instance, sound will travel 1.59 times faster in nickel than in bronze, due to the greater stiffness of nickel at about the same density. Similarly, sound travels about 1.41 times faster in light hydrogen (protium) gas than in heavy hydrogen (deuterium) gas, since deuterium has similar properties but twice the density. At the same time, "compression-type" sound will travel faster in solids than in liquids, and faster in liquids than in gases, because the solids are more difficult to compress than liquids, while liquids in turn are more difficult to compress than gases.

Some textbooks mistakenly state that the speed of sound increases with increasing density. This is usually illustrated by presenting data for three materials, such as air, water and steel, which also have vastly different compressibilities which more than make up for the density differences. An illustrative example of the two effects is that sound travels only 4.3 times faster in water than air, despite enormous differences in compressibility of the two media. The reason is that the larger density of water, which works to *slow* sound in water relative to air, nearly makes up for the compressibility differences in the two media.

Compression and Shear Waves

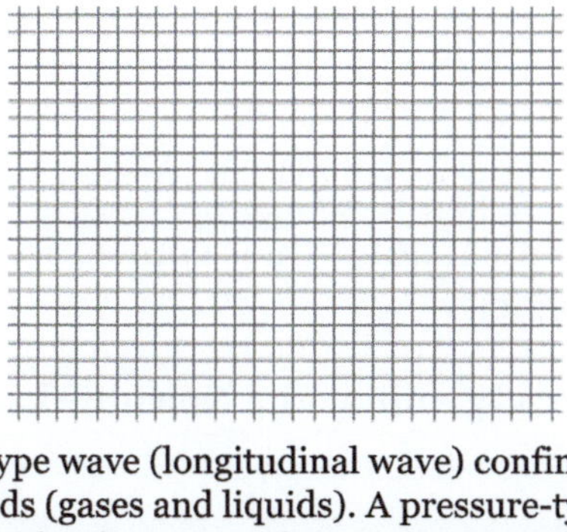

Pressure-pulse or compression-type wave (longitudinal wave) confined to a plane. This is the only type of sound wave that travels in fluids (gases and liquids). A pressure-type wave may also travel in solids, along with other types of waves (transverse waves)

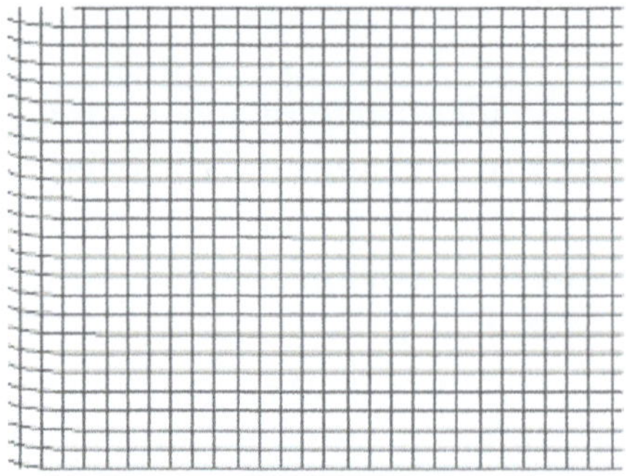

Transverse wave affecting atoms initially confined to a plane. This additional type of sound wave (additional type of elastic wave) travels only in solids, for it requires a sideways shearing motion which is supported by the presence of elasticity in the solid. The sideways shearing motion may take place in **any** direction which is at right-angle to the direction of wave-travel (only one shear direction is shown here, at right angles to the plane). Furthermore, the right-angle shear direction may change over time and distance, resulting in different types of *polarization* of shear-waves

In a gas or liquid, sound consists of compression waves. In solids, waves propagate as two different types. A longitudinal wave is associated with compression and decompression in the direction of travel, and is the same process in gases and liquids, with an analogous compression-type wave in solids. Only compression waves are supported in gases and liquids. An additional type of wave, the transverse wave, also called a shear wave, occurs only in solids because only solids support elastic deformations. It is due to elastic deformation of the medium perpendicular to the direction of wave travel; the direction of shear-deformation is called the "polarization" of this type of wave. In general, transverse waves occur as a pair of orthogonal polarizations.

These different waves (compression waves and the different polarizations of shear waves) may have different speeds at the same frequency. Therefore, they arrive at an observer at different times, an extreme example being an earthquake, where sharp compression waves arrive first, and rocking transverse waves seconds later.

The speed of a compression wave in fluid is determined by the medium's compressibility and density. In solids, the compression waves are analogous to those in fluids, depending on compressibility, density, and the additional factor of shear modulus. The speed of shear waves, which can occur only in solids, is determined simply by the solid material's shear modulus and density.

Equations

The speed of sound in mathematical notation is conventionally represented by *c*, from the Latin *celeritas* meaning "velocity".

In general, the speed of sound *c* is given by the Newton–Laplace equation:

$$c = \sqrt{\frac{K_s}{\rho}},$$

where

- K_s is a coefficient of stiffness, the isentropic bulk modulus (or the modulus of bulk elasticity for gases);
- ρ is the density.

Thus the speed of sound increases with the stiffness (the resistance of an elastic body to deformation by an applied force) of the material, and decreases with the density. For ideal gases the bulk modulus K is simply the gas pressure multiplied by the dimensionless adiabatic index, which is about 1.4 for air under normal conditions of pressure and temperature.

For general equations of state, if classical mechanics is used, the speed of sound c is given by

$$c = \sqrt{\left(\frac{\partial p}{\partial \rho}\right)_s},$$

where

- p is the pressure;
- ρ is the density and the derivative is taken isentropically, that is, at constant entropy s.

If relativistic effects are important, the speed of sound is calculated from the relativistic Euler equations.

In a non-dispersive medium, the speed of sound is independent of sound frequency, so the speeds of energy transport and sound propagation are the same for all frequencies. Air, a mixture of oxygen and nitrogen, constitutes a non-dispersive medium. However, air does contain a small amount of CO_2 which *is* a dispersive medium, and causes dispersion to air at ultrasonic frequencies (> 28 kHz).

In a dispersive medium, the speed of sound is a function of sound frequency, through the dispersion relation. Each frequency component propagates at its own speed, called the phase velocity, while the energy of the disturbance propagates at the group velocity. The same phenomenon occurs with light waves.

Dependence on the Properties of the Medium

The speed of sound is variable and depends on the properties of the substance through which the wave is travelling. In solids, the speed of transverse (or shear) waves depends on the shear deformation under shear stress (called the shear modulus), and the density of the medium. Longitudinal (or compression) waves in solids depend on the same two factors with the addition of a dependence on compressibility.

In fluids, only the medium's compressibility and density are the important factors, since fluids do not transmit shear stresses. In heterogeneous fluids, such as a liquid filled with gas bubbles, the density of the liquid and the compressibility of the gas affect the speed of sound in an additive manner, as demonstrated in the hot chocolate effect.

In gases, adiabatic compressibility is directly related to pressure through the heat capacity ratio (adiabatic index), while pressure and density are inversely related to the temperature and molecular weight, thus making only the completely independent properties of *temperature and molecular structure* important (heat capacity ratio may be determined by temperature and molecular structure, but simple molecular weight is not sufficient to determine it).

In low molecular weight gases such as helium, sound propagates faster as compared to heavier gases such as xenon. For monatomic gases, the speed of sound is about 75% of the mean speed that the atoms move in that gas.

For a given ideal gas the molecular composition is fixed, and thus the speed of sound depends only on its temperature. At a constant temperature, the gas pressure has no effect on the speed of sound, since the density will increase, and since pressure and density (also proportional to pressure) have equal but opposite effects on the speed of sound, and the two contributions cancel out exactly. In a similar way, compression waves in solids depend both on compressibility and density—just as in liquids—but in gases the density contributes to the compressibility in such a way that some part of each attribute factors out, leaving only a dependence on temperature, molecular weight, and heat capacity ratio which can be independently derived from temperature and molecular composition. Thus, for a single given gas (assuming the molecular weight does not change) and over a small temperature range (for which the heat capacity is relatively constant), the speed of sound becomes dependent on only the temperature of the gas.

In non-ideal gas behavior regimen, for which the van der Waals gas equation would be used, the proportionality is not exact, and there is a slight dependence of sound velocity on the gas pressure.

Humidity has a small but measurable effect on the speed of sound (causing it to increase by about 0.1%–0.6%), because oxygen and nitrogen molecules of the air are replaced by lighter molecules of water. This is a simple mixing effect.

Altitude variation and Implications for Atmospheric Acoustics

In the Earth's atmosphere, the chief factor affecting the speed of sound is the temperature. For a given ideal gas with constant heat capacity and composition, the speed of sound is dependent *solely* upon temperature. In such an ideal case, the effects of decreased density and decreased pressure of altitude cancel each other out, save for the residual effect of temperature.

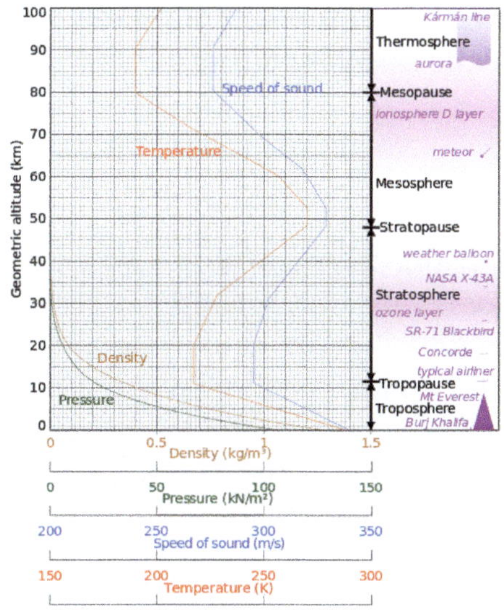

Density and pressure decrease smoothly with altitude, but temperature (red) does not. The speed of sound (blue) depends only on the complicated temperature variation at altitude and can be calculated from it, since isolated density and pressure effects on the speed of sound cancel each other. Speed of sound increases with height in two regions of the stratosphere and thermosphere, due to heating effects in these regions.

Since temperature (and thus the speed of sound) decreases with increasing altitude up to 11 km, sound is refracted upward, away from listeners on the ground, creating an acoustic shadow at some distance from the source. The decrease of the speed of sound with height is referred to as a negative sound speed gradient.

However, there are variations in this trend above 11 km. In particular, in the stratosphere above about 20 km, the speed of sound increases with height, due to an increase in temperature from heating within the ozone layer. This produces a positive speed of sound gradient in this region. Still another region of positive gradient occurs at very high altitudes, in the aptly-named thermosphere above 90 km.

Practical Formula for Dry Air

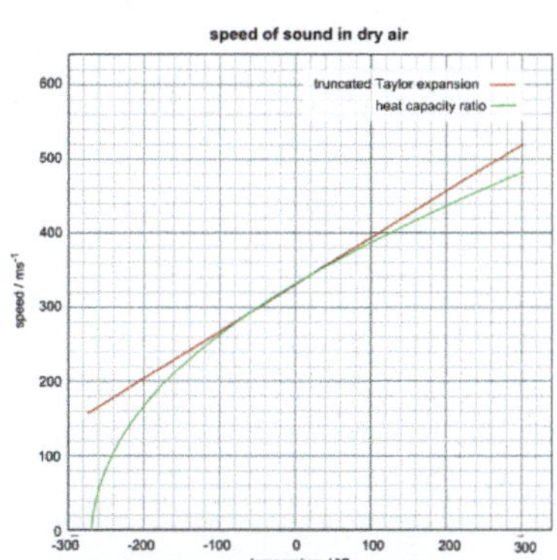

Approximation of the speed of sound in dry air based on the heat capacity ratio (in green) against the truncated Taylor expansion (in red).

The approximate speed of sound in dry (0% humidity) air, in meters per second, at temperatures near 0 °C, can be calculated from

$$c_{\mathrm{air}} = (331.3 + 0.606 \cdot \vartheta)\ \mathrm{m/s},$$

where ϑ is the temperature in degrees Celsius (°C).

This equation is derived from the first two terms of the Taylor expansion of the following more accurate equation:

$$c_{\mathrm{air}} = 331.3\sqrt{1 + \frac{\vartheta}{273.15}}\ \mathrm{m/s}.$$

Dividing the first part, and multiplying the second part, on the right hand side, by √273.15 gives the exactly equivalent form

$$c_{\mathrm{air}} = 20.05\sqrt{\vartheta + 273.15}\ \mathrm{m/s}.$$

The value of 331.3 m/s, which represents the speed at 0 °C (or 273.15 K), is based on theoretical (and some measured) values of the heat capacity ratio, γ, as well as on the fact that at 1 atm real air is very well described by the ideal gas approximation. Commonly found values for the speed of sound at 0 °C may vary from 331.2 to 331.6 due to the assumptions made when it is calculated. If ideal gas γ is assumed to be 7/5 = 1.4 exactly, the 0 °C speed is calculated to be 331.3 m/s, the coefficient used above.

This equation is correct to a much wider temperature range, but still depends on the approximation of heat capacity ratio being independent of temperature, and for this reason will fail, particularly at higher temperatures. It gives good predictions in relatively dry, cold, low pressure conditions, such as the Earth's stratosphere. The equation fails at extremely low pressures and short wavelengths, due to dependence on the assumption that the wavelength of the sound in the gas is much longer than the average mean free path between gas molecule collisions. A derivation of these equations will be given in the following section.

A graph comparing results of the two equations is at right, using the slightly different value of 331.5 m/s for the speed of sound at 0 °C.

Details

Speed of Sound in Ideal Gases and Air

For an ideal gas, K (the bulk modulus in equations above, equivalent to C, the coefficient of stiffness in solids) is given by

$$K = \gamma \cdot p,$$

thus, from the Newton–Laplace equation above, the speed of sound in an ideal gas is given by

$$c = \sqrt{\gamma \cdot \frac{p}{\rho}},$$

where

- γ is the adiabatic index also known as the *isentropic expansion factor*. It is the ratio of specific heats of a gas at a constant-pressure to a gas at a constant-volume(), and arises because a classical sound wave induces an adiabatic compression, in which the heat of the compression does not have enough time to escape the pressure pulse, and thus contributes to the pressure induced by the compression;
- p is the pressure;
- ρ is the density.

Using the ideal gas law to replace p with nRT/V, and replacing ρ with nM/V, the equation for an ideal gas becomes

$$c_{\text{ideal}} = \sqrt{\gamma \cdot \frac{p}{\rho}} = \sqrt{\frac{\gamma \cdot R \cdot T}{M}} = \sqrt{\frac{\gamma \cdot k \cdot T}{m}},$$

where

- c_{ideal} is the speed of sound in an ideal gas;
- R (approximately 8.314,5 $\text{J} \cdot \text{mol}^{-1} \cdot \text{K}^{-1}$) is the molar gas constant(universal gas constant);
- k is the Boltzmann constant;
- γ (gamma) is the adiabatic index. At room temperature, where thermal energy is fully partitioned into rotation (rotations are fully excited) but quantum effects prevent excitation of vibrational modes, the value is 7/5 = 1.400 for diatomic molecules, according to kinetic theory. Gamma is actually experimentally measured over a range from 1.399,1 to 1.403 at 0 °C, for air. Gamma is exactly 5/3 = 1.6667 for monatomic gases such as noble gases;
- T is the absolute temperature;
- M is the molar mass of the gas. The mean molar mass for dry air is about 0.028,964,5 kg/mol;
- n is the number of moles;
- m is the mass of a single molecule.

This equation applies only when the sound wave is a small perturbation on the ambient condition, and the certain other noted conditions are fulfilled, as noted below. Calculated values for c_{air} have been found to vary slightly from experimentally determined values.

Newton famously considered the speed of sound before most of the development of thermodynamics and so incorrectly used isothermal calculations instead of adiabatic. His result was missing the factor of γ but was otherwise correct.

Numerical substitution of the above values gives the ideal gas approximation of sound velocity for gases, which is accurate at relatively low gas pressures and densities (for air, this includes standard Earth sea-level conditions). Also, for diatomic gases the use of $\gamma = 1.400{,}0$ requires that the gas exists in a temperature range high enough that rotational heat capacity is fully excited (i.e., molecular rotation is fully used as a heat energy "partition" or reservoir); but at the same time the temperature must be low enough that molecular vibrational modes contribute no heat capacity (i.e., insignificant heat goes into vibration, as all vibrational quantum modes above the minimum-energy-mode, have energies too high to be populated by a significant number of molecules at this temperature). For air, these conditions are fulfilled at room temperature, and also temperatures considerably below room temperature.

For air, we use a simplified symbol

$$R_* = R / M_{air}.$$

Additionally, if temperatures in degrees Celsius(°C) are to be used to calculate air speed in the region near 273 kelvin, then Celsius temperature $\theta = T - 273.15$ may be used. Then

$$c_{ideal} = \sqrt{\gamma \cdot R_* \cdot T} = \sqrt{\gamma \cdot R_* \cdot (\vartheta + 273.15)},$$

$$c_{ideal} = \sqrt{\gamma \cdot R_* \cdot 273.15} \cdot \sqrt{1 + \frac{\vartheta}{273.15}}.$$

For dry air, where θ (theta) is the temperature in degrees Celsius(°C).

Making the following numerical substitutions,

$$R = 8.314{,}510\,\text{J/(mol} \cdot \text{K)}$$

is the molar gas constant in J/mole/Kelvin, and

$$M_{air} = 0.028{,}964{,}5\,\text{kg/mol}$$

is the mean molar mass of air, in kg; and using the ideal diatomic gas value of $\gamma = 1.4000$.

Then

$$c_{\text{air}} = 331.3 \sqrt{1 + \frac{\vartheta^{\circ}\text{C}}{273.15^{\circ}\text{C}}} \ \ \text{m/s}.$$

Using the first two terms of the Taylor expansion:

$$c_{\text{air}} = 331.3 \ (1 + \frac{\vartheta^{\circ}\text{C}}{2 \cdot 273.15^{\circ}\text{C}}) \ \ \text{m/s},$$

$$c_{\text{air}} = (331.3 + 0.606^{\circ}\text{C}^{-1} \cdot \vartheta) \ \ \text{m/s}.$$

The derivation includes the first two equations given in the *Practical formula for dry* air section above.

Effects Due to Wind Shear

The speed of sound varies with temperature. Since temperature and sound velocity normally decrease with increasing altitude, sound is refracted upward, away from listeners on the ground, creating an acoustic shadow at some distance from the source. Wind shear of 4 m/(s · km) can produce refraction equal to a typical temperature lapse rate of 7.5 °C/km. Higher values of wind gradient will refract sound downward toward the surface in the downwind direction, eliminating the acoustic shadow on the downwind side. This will increase the audibility of sounds downwind. This downwind refraction effect occurs because there is a wind gradient; the sound is not being carried along by the wind.

For sound propagation, the exponential variation of wind speed with height can be defined as follows:

$$U(h) = U(0)h^{\zeta},$$

$$\frac{\mathrm{d}U}{\mathrm{d}H}(h) = \zeta \frac{U(h)}{h},$$

where

- $U(h)$ is the speed of the wind at height h;
- ζ is the exponential coefficient based on ground surface roughness, typically between 0.08 and 0.52;
- $\mathrm{d}U/\mathrm{d}H(h)$ is the expected wind gradient at height h.

In the 1862 American Civil War Battle of Iuka, an acoustic shadow, believed to have been enhanced by a northeast wind, kept two divisions of Union soldiers out of the battle, because they could not hear the sounds of battle only 10 km (six miles) downwind.

Tables

In the standard atmosphere:

- T_0 is 273.15 K (= 0 °C = 32 °F), giving a theoretical value of 331.3 m/s (= 1086.9 ft/s = 1193 km/h = 741.1 mph = 644.0 kn). Values ranging from 331.3-331.6 may be found in reference literature, however;
- T_{20} is 293.15 K (= 20 °C = 68 °F), giving a value of 343.2 m/s (= 1126.0 ft/s = 1236 km/h = 767.8 mph = 667.2 kn);
- T_{25} is 298.15 K (= 25 °C = 77 °F), giving a value of 346.1 m/s (= 1135.6 ft/s = 1246 km/h = 774.3 mph = 672.8 kn).

In fact, assuming an ideal gas, the speed of sound c depends on temperature only, not on the pressure or density (since these change in lockstep for a given temperature and cancel out). Air is almost an ideal gas. The temperature of the air varies with altitude, giving the following variations in the speed of sound using the standard atmosphere—*actual conditions may vary.*

Effect of temperature on properties of air			
Temperature T (°C)	**Speed of sound c (m/s)**	**Density of air ρ (kg/m³)**	**Characteristic specific acoustic impedance z_0 (Pa·s/m)**
35	351.88	1.1455	403.2
30	349.02	1.1644	406.5
25	346.13	1.1839	409.4
20	343.21	1.2041	413.3
15	340.27	1.2250	416.9
10	337.31	1.2466	420.5
5	334.32	1.2690	424.3
0	331.30	1.2922	428.0
−5	328.25	1.3163	432.1
−10	325.18	1.3413	436.1
−15	322.07	1.3673	440.3
−20	318.94	1.3943	444.6
−25	315.77	1.4224	449.1

Given normal atmospheric conditions, the temperature, and thus speed of sound, varies with altitude:

Altitude	Temperature	m/s	km/h	mph	kn
Sea level	15 °C (59 °F)	340	1,225	761	661
11,000 m–20,000 m (Cruising altitude of commercial jets, and first supersonic flight)	−57 °C (−70 °F)	295	1,062	660	573
29,000 m (Flight of X-43A)	−48 °C (−53 °F)	301	1,083	673	585

Effect of Frequency and Gas Composition

General Physical Considerations

The medium in which a sound wave is travelling does not always respond adiabatically, and as a result the speed of sound can vary with frequency.

The limitations of the concept of speed of sound due to extreme attenuation are also of concern. The attenuation which exists at sea level for high frequencies applies to successively lower frequencies as atmospheric pressure decreases, or as the mean free path increases. For this reason, the concept of speed of sound (except for frequencies approaching zero) progressively loses its range of applicability at high altitudes. The standard equations for the speed of sound apply with reasonable accuracy only to situations in which the wavelength of the soundwave is considerably longer than the mean free path of molecules in a gas.

The molecular composition of the gas contributes both as the mass (M) of the molecules, and their heat capacities, and so both have an influence on speed of sound. In general, at the same molecular mass, monatomic gases have slightly higher speed of sound (over 9% higher) because they have a higher γ (5/3 = 1.66...) than diatomics do (7/5 = 1.4). Thus, at the same molecular mass, the speed of sound of a monatomic gas goes up by a factor of

$$\frac{c_{\text{gas,monatomic}}}{c_{\text{gas,diatomic}}} = \sqrt{\frac{5/3}{7/5}} = \sqrt{\frac{25}{21}} = 1.091\ldots$$

This gives the 9% difference, and would be a typical ratio for speeds of sound at room temperature in helium vs. deuterium, each with a molecular weight of 4. Sound travels faster in helium than deuterium because adiabatic compression heats helium more, since the helium molecules can store heat energy from compression only in translation, but not rotation. Thus helium molecules (monatomic molecules) travel faster in a sound wave and transmit sound faster. (Sound travels at about 70% of the mean molecular speed in gases; the figure is 75% in monatomic gases and 68% in diatomic gases).

Note that in this example we have assumed that temperature is low enough that heat capacities are not influenced by molecular vibration. However, vibrational modes simply cause gammas which decrease toward 1, since vibration modes in a polyatomic gas gives the gas additional ways to store heat which do not affect temperature, and thus do not affect molecular velocity and sound velocity. Thus, the effect of higher temperatures and vibrational heat capacity acts to increase the difference between the speed of sound in monatomic vs. polyatomic molecules, with the speed remaining greater in monatomics.

Practical Application to Air

By far the most important factor influencing the speed of sound in air is temperature. The speed is proportional to the square root of the absolute temperature, giving an increase of about 0.6 m/s per degree Celsius. For this reason, the pitch of a musical wind instrument increases as its temperature increases.

The speed of sound is raised by humidity but decreased by carbon dioxide. The difference between 0% and 100% humidity is about 1.5 m/s at standard pressure and temperature, but the size of the humidity effect increases dramatically with temperature. The carbon dioxide content of air is not fixed, due to both carbon pollution and human breath (e.g., in the air blown through wind instruments).

The dependence on frequency and pressure are normally insignificant in practical applications. In dry air, the speed of sound increases by about 0.1 m/s as the frequency rises from 10 Hz to 100 Hz. For audible frequencies above 100 Hz it is relatively constant. Standard values of the speed of sound are quoted in the limit of low frequencies, where the wavelength is large compared to the mean free path.

Mach Number

Mach number, a useful quantity in aerodynamics, is the ratio of air speed to the local speed of sound. At altitude, for reasons explained, Mach number is a function of temperature. Aircraft flight instruments, however, operate using pressure differential to compute Mach number, not temperature. The assumption is that a particular pressure represents a particular altitude and, therefore, a standard temperature. Aircraft flight instruments need to operate this way because the stagnation pressure sensed by a Pitot tube is dependent on altitude as well as speed.

Experimental Methods

A range of different methods exist for the measurement of sound in air.

The earliest reasonably accurate estimate of the speed of sound in air was made by William Derham, and acknowledged by Isaac Newton. Derham had a telescope at the top of the tower of the Church of St Laurence in Upminster, England. On a calm day, a synchronized pocket watch would be given to an assistant who would fire a shotgun at a pre-determined time from a conspicuous point some miles away, across the countryside. This could be confirmed by telescope. He then measured the interval between seeing gunsmoke and arrival of the sound using a half-second pendulum. The distance from where the gun was fired was found by triangulation, and simple division (distance/time) provided velocity. Lastly, by making many observations, using a range of different distances, the inaccuracy of the half-second pendulum could be averaged out, giving his final estimate of the speed of sound. Modern stopwatches enable this method to be used today over distances as short as 200–400 meters, and not needing something as loud as a shotgun.

Single-shot Timing Methods

The simplest concept is the measurement made using two microphones and a fast recording device such as a digital storage scope. This method uses the following idea.

If a sound source and two microphones are arranged in a straight line, with the sound source at one end, then the following can be measured:

1. The distance between the microphones (x), called microphone basis.
2. The time of arrival between the signals (delay) reaching the different microphones (t).

Then $v = x/t$.

Other Methods

In these methods the time measurement has been replaced by a measurement of the inverse of time (frequency).

Kundt's tube is an example of an experiment which can be used to measure the speed of sound in a small volume. It has the advantage of being able to measure the speed of sound in any gas. This method uses a powder to make the nodes and antinodes visible to the human eye. This is an example of a compact experimental setup.

A tuning fork can be held near the mouth of a long pipe which is dipping into a barrel of water. In this system it is the case that the pipe can be brought to resonance if the length of the air column in the pipe is equal to $(1 + 2n)\lambda/4$ where n is an integer. As the antinodal point for the pipe at the open end is slightly outside the mouth of the pipe it is best to find two or more points of resonance and then measure half a wavelength between these.

Here it is the case that $v = f\lambda$.

High-precision Measurements in Air

The effect from impurities can be significant when making high-precision measurements. Chemical desiccants can be used to dry the air, but will in turn contaminate the sample. The air can be dried cryogenically, but this has the effect of removing the carbon dioxide as well; therefore many high-precision measurements are performed with air free of carbon dioxide rather than with natural air. A 2002 review found that a 1963 measurement by Smith and Harlow using a cylindrical resonator gave "the most probable value of the standard speed of sound to date." The experiment was done with air from which the carbon dioxide had been removed, but the result was then corrected for this effect so as to be applicable to real air. The experiments were done at 30 °C but corrected for temperature in order to report them at 0 °C. The result was 331.45 ± 0.01 m/s for dry air at STP, for frequencies from 93 Hz to 1,500 Hz.

Non-gaseous Media

Three-dimensional Solids

In a solid, there is a non-zero stiffness both for volumetric deformations and shear deformations. Hence, it is possible to generate sound waves with different velocities dependent on the deformation mode. Sound waves generating volumetric deformations (compression) and shear deformations (shearing) are called pressure waves (longitudinal waves) and shear waves (transverse waves), respectively. In earthquakes, the corresponding seismic waves are called P-waves (primary waves) and S-waves (secondary waves), respectively. The sound velocities of these two types of waves propagating in a homogeneous 3-dimensional solid are respectively given by

$$c_{\mathrm{solid,p}} = \sqrt{\frac{K + \frac{4}{3}G}{\rho}} = \sqrt{\frac{E(1-\nu)}{\rho(1+\nu)(1-2\nu)}},$$

$$c_{\mathrm{solid,s}} = \sqrt{\frac{G}{\rho}},$$

where

- K is the bulk modulus of the elastic materials;
- G is the shear modulus of the elastic materials;
- E is the Young's modulus;
- ρ is the density;
- ν is Poisson's ratio.

The last quantity is not an independent one, as $E = 3K(1 - 2\nu)$. Note that the speed of pressure waves depends both on the pressure and shear resistance properties of the material, while the speed of shear waves depends on the shear properties only.

Typically, pressure waves travel faster in materials than do shear waves, and in earthquakes this is the reason that the onset of an earthquake is often preceded by a quick upward-downward shock, before arrival of waves that produce a side-to-side motion. For example, for a typical steel alloy, $K = 170$ GPa, $G = 80$ GPa and $\rho = 7{,}700$ kg/m^3, yielding a compressional speed $c_{\text{solid,p}}$ of 6,000 m/s. This is in reasonable agreement with $c_{\text{solid,p}}$ measured experimentally at 5,930 m/s for a (possibly different) type of steel. The shear speed $c_{\text{solid,s}}$ is estimated at 3,200 m/s using the same numbers.

One-dimensional Solids

The speed of sound for pressure waves in stiff materials such as metals is sometimes given for "long rods" of the material in question, in which the speed is easier to measure. In rods where their diameter is shorter than a wavelength, the speed of pure pressure waves may be simplified and is given by:

$$c_{\text{solid}} = \sqrt{\frac{E}{\rho}},$$

where E is the Young's modulus. This is similar to the expression for shear waves, save that Young's modulus replaces the shear modulus. This speed of sound for pressure waves in long rods will always be slightly less than the same speed in homogeneous 3-dimensional solids, and the ratio of the speeds in the two different types of objects depends on Poisson's ratio for the material.

Speed of Sound in Liquids

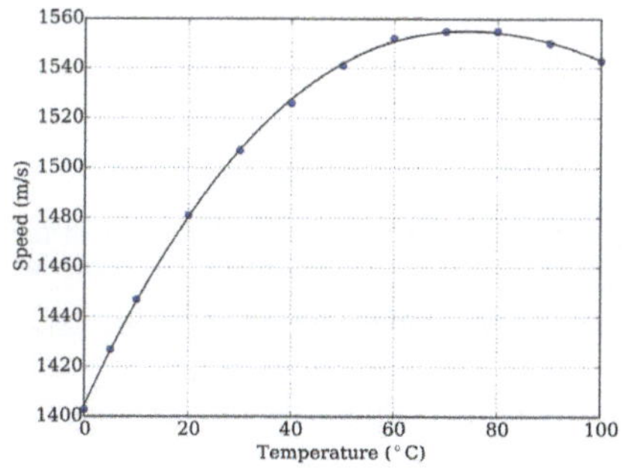

Speed of sound in water vs temperature.

In a fluid the only non-zero stiffness is to volumetric deformation (a fluid does not sustain shear forces).

Hence the speed of sound in a fluid is given by

$$c_{\text{fluid}} = \sqrt{\frac{K}{\rho}},$$

where K is the bulk modulus of the fluid.

Water

In fresh water, sound travels at about 1481 m/s at 20 °C. Speed of Sound in Pure Water for an online calculator. Applications of underwater sound can be found in sonar, acoustic communication and acoustical oceanography.

Seawater

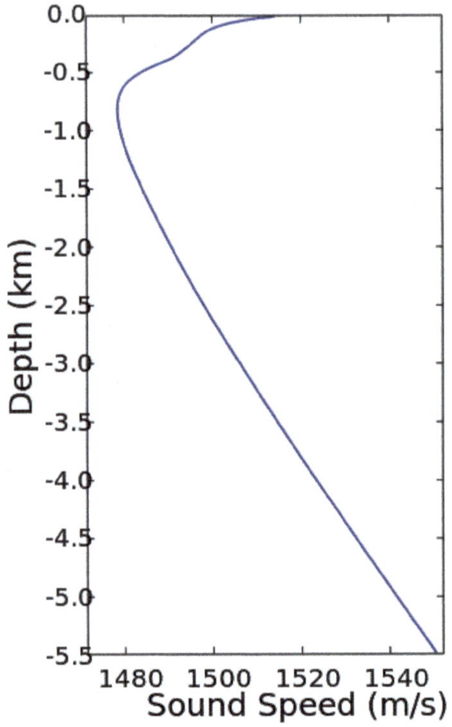

Speed of sound as a function of depth at a position north of Hawaii in the Pacific Ocean derived from the 2005 World Ocean Atlas. The SOFAR channel is centred on the minimum in the speed of sound at ca. 750-m depth.

In salt water that is free of air bubbles or suspended sediment, sound travels at about 1500 m/s (1500.235 m/s at 1000 kilopascals, 10 °C and 3% salinity by one method). The speed of sound in seawater depends on pressure (hence depth), temperature (a change of 1 °C ~ 4 m/s), and salinity (a change of 1‰ ~ 1 m/s), and empirical equations have been derived to accurately calculate the speed of sound from these variables. Other factors affecting the speed of sound are minor. Since temperature decreases with depth while pressure and generally salinity increase, the profile of the speed of sound with depth generally shows a characteristic curve which decreases to a minimum at a depth of several hundred meters, then increases again with increasing depth (right). For more information.

A simple empirical equation for the speed of sound in sea water with reasonable accuracy for the world's oceans is due to Mackenzie:

$$c(T,S,z) = a_1 + a_2T + a_3T^2 + a_4T^3 + a_5(S-35) + a_6z + a_7z^2 + a_8T(S-35) + a_9Tz^3,$$

where

- T is the temperature in degrees Celsius;
- S is the salinity in parts per thousand;
- z is the depth in meters.

The constants $a_1, a_2, \ldots, a_9$ are

$$\begin{array}{llllll} a_1 & = 1,448.96, & a_2 & = 4.591, & a_3 & = -5.304\times10^{-2}, \\ a_4 & = 2.374\times10^{-4}, & a_5 & = 1.340, & a_6 & = 1.630\times10^{-2}, \\ a_7 & = 1.675\times10^{-7}, & a_8 & = -1.025\times10^{-2}, & a_9 & = -7.139\times10^{-13}, \end{array}$$

with check value 1550.744 m/s for T = 25 °C, S = 35 parts per thousand, z = 1,000 m. This equation has a standard error of 0.070 m/s for salinity between 25 and 40 ppt.

Other equations for the speed of sound in sea water are accurate over a wide range of conditions, but are far more complicated, e.g., that by V. A. Del Grosso and the Chen-Millero-Li Equation.

Speed of Sound in Plasma

The speed of sound in a plasma for the common case that the electrons are hotter than the ions (but not too much hotter) is given by the formula.

$$c_s = (\gamma ZkT_e / m_i)^{1/2} = 9.79\times10^3(\gamma ZT_e / \mu)^{1/2}\ \text{m/s},$$

where

- m_i is the ion mass;
- μ is the ratio of ion mass to proton mass $\mu = m_i/m_p$;
- T_e is the electron temperature;
- Z is the charge state;
- k is Boltzmann constant;
- γ is the adiabatic index.

In contrast to a gas, the pressure and the density are provided by separate species, the pressure by the electrons and the density by the ions. The two are coupled through a fluctuating electric field.

Gradients

When sound spreads out evenly in all directions in three dimensions, the intensity drops in proportion to the inverse square of the distance. However, in the ocean there is a layer called the 'deep sound channel' or SOFAR channel which can confine sound waves at a particular depth.

In the SOFAR channel, the speed of sound is lower than that in the layers above and below. Just as light waves will refract towards a region of higher index, sound waves will refract towards a region where their speed is reduced. The result is that sound gets confined in the layer, much the way light can be confined in a sheet of glass or optical fiber. Thus, the sound is confined in essentially two dimensions. In two dimensions the intensity drops in proportion to only the inverse of the distance. This allows waves to travel much further before being undetectably faint.

A similar effect occurs in the atmosphere. Project Mogul successfully used this effect to detect a nuclear explosion at a considerable distance.

Amplitude

The amplitude of a periodic variable is a measure of its change over a single period (such as time or spatial period). There are various definitions of amplitude, which are all functions of the magnitude of the difference between the variable's extreme values. In older texts the phase is sometimes called the amplitude.

Definitions

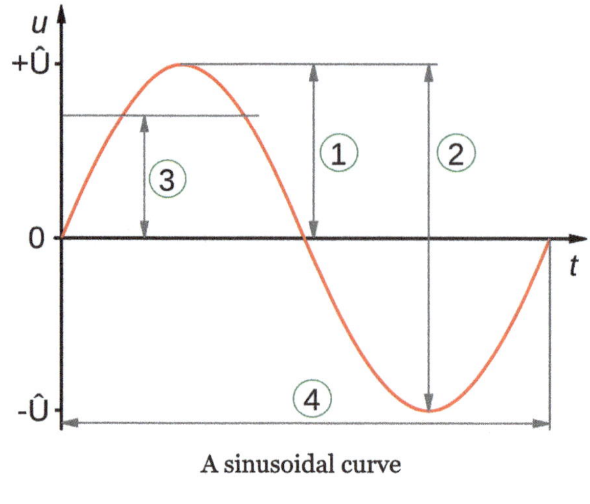

A sinusoidal curve

1. Peak amplitude ($\hat{U}$),
2. Peak-to-peak amplitude ($2\hat{U}$),
3. Root mean square amplitude ($\hat{U} / \sqrt{2}$),
4. Wave period (not an amplitude)

Peak-to-peak Amplitude

Peak-to-peak amplitude is the change between peak (highest amplitude value) and trough (lowest amplitude value, which can be negative). With appropriate circuitry, peak-to-peak amplitudes of electric oscillations can be measured by meters or by viewing the waveform on an oscilloscope. Peak-to-peak is a straightforward measurement on an oscilloscope, the peaks of the waveform being easily identified and measured against the graticule. This remains a common way of specifying amplitude, but sometimes other measures of amplitude are more appropriate.

Peak Amplitude

In audio system measurements, telecommunications and other areas where the measurand is a signal that swings above and below a reference value but is not sinusoidal, peak amplitude is often used. If the reference is zero, this is the maximum absolute value of the signal; if the reference is a mean value (DC component), the peak amplitude is the maximum absolute value of the difference from that reference.

Semi-amplitude

Semi-amplitude means half the peak-to-peak amplitude. Some scientists use *amplitude* or *peak amplitude* to mean semi-amplitude, that is, half the peak-to-peak amplitude.

It is the most widely used measure of orbital wobble in astronomy and the measurement of small radial velocity semi-amplitudes of nearby stars is important in the search for exoplanets.

Root Mean Square Amplitude

Root mean square (RMS) amplitude is used especially in electrical engineering: the RMS is defined as the square root of the mean over time of the square of the vertical distance of the graph from the rest state, i.e., the RMS of the AC waveform (with no DC component).

For complicated waveforms, especially non-repeating signals like noise, the RMS amplitude is usually used because it is both unambiguous and has physical significance. For example, the *average* power transmitted by an acoustic or electromagnetic wave or by an electrical signal is proportional to the square of the RMS amplitude (and not, in general, to the square of the peak amplitude).

For alternating current electric power, the universal practice is to specify RMS values of a sinusoidal waveform. One property of root mean square voltages and currents is that they produce the same heating effect as direct current in a given resistance.

The peak-to-peak value is used, for example, when choosing rectifiers for power supplies, or when estimating the maximum voltage that insulation must withstand. Some common voltmeters are calibrated for RMS amplitude, but respond to the average value of a rectified waveform. Many digital voltmeters and all moving coil meters are in this category. The RMS calibration is only correct for a sine wave input since the ratio between peak, average and RMS values is dependent on waveform. If the wave shape being measured is greatly different from a sine wave, the relationship between RMS and average value changes. True RMS-responding meters were used in radio frequency measurements, where instruments measured the heating effect in a resistor to measure current. The advent of microprocessor controlled meters capable of calculating RMS by sampling the waveform has made true RMS measurement commonplace.

Ambiguity

In general, the use of peak amplitude is simple and unambiguous only for symmetric periodic waves, like a sine wave, a square wave, or a triangular wave. For an asymmetric wave (periodic pulses in one direction, for example), the peak amplitude becomes ambiguous. This is because the value is different depending on whether the maximum positive signal is measured relative to the mean, the maximum negative signal is measured relative to the mean, or the maximum positive signal is measured relative to the maximum negative signal (the *peak-to-peak amplitude*) and then divided by two. In electrical engineering, the usual solution to this ambiguity is to measure the amplitude from a defined reference potential (such as ground or 0 V). Strictly speaking, this is no longer amplitude since there is the possibility that a constant (DC component) is included in the measurement.

Pulse Amplitude

In telecommunication, *pulse amplitude* is the magnitude of a pulse parameter, such as the voltage level, current level, field intensity, or power level.

Pulse amplitude is measured with respect to a specified reference and therefore should be modified by qualifiers, such as *average*, *instantaneous*, *peak*, or *root-mean-square*.

Pulse amplitude also applies to the amplitude of frequency- and phase-modulated waveform envelopes.

Formal Representation

In this simple wave equation

$$x = A\sin(\omega[t - K]) + b\,,$$

- A is the amplitude (or peak amplitude),
- x is the oscillating variable,
- ω is angular frequency,
- t is time,
- K and b are arbitrary constants representing time and displacement offsets respectively.

Units

The units of the amplitude depend on the type of wave, but are always in the same units as the oscillating variable. A more general representation of the wave equation is more complex, but the role of amplitude remains analogous to this simple case.

For waves on a string, or in medium such as water, the amplitude is a displacement.

The amplitude of sound waves and audio signals (which relates to the volume) conventionally refers to the amplitude of the air pressure in the wave, but sometimes the amplitude of the displacement (movements of the air or the diaphragm of a speaker) is described. The logarithm of the amplitude squared is usually quoted in dB, so a null amplitude corresponds to $-\infty$ dB. Loudness is related to amplitude and intensity and is one of the most salient qualities of a sound, although in general sounds can be recognized independently of amplitude. The square of the amplitude is proportional to the intensity of the wave.

For electromagnetic radiation, the amplitude of a photon corresponds to the changes in the electric field of the wave. However, radio signals may be carried by electromagnetic radiation; the intensity of the radiation (amplitude modulation) or the frequency of the radiation (frequency modulation) is oscillated and then the individual oscillations are varied (modulated) to produce the signal.

Waveform and Envelope

The amplitude as defined above is a constant and the wave is said to be wavelength continuous. If this condition does not hold, amplitude alike variations with time and/or position may be quantified in terms of the envelope of the wave.

Wavelength

In physics, the wavelength of a sinusoidal wave is the spatial period of the wave—the distance over which the wave's shape repeats, and thus the inverse of the spatial frequency. It is usually determined by considering the distance between consecutive corresponding points of the same phase, such as crests, troughs, or zero crossings and is a characteristic of both traveling waves and standing waves, as well as other spatial wave patterns. Wavelength is commonly designated by the Greek letter *lambda* (λ). The concept can also be applied to periodic waves of non-sinusoidal shape. The term *wavelength* is also sometimes applied to modulated waves, and to the sinusoidal envelopes of modulated waves or waves formed by interference of several sinusoids.

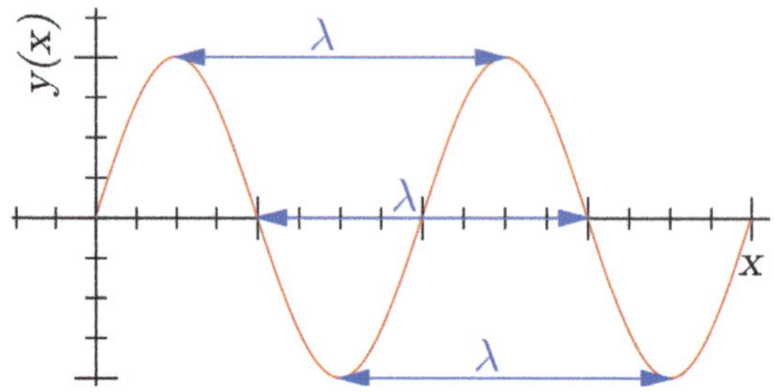

Wavelength of a sine wave, λ, can be measured between any two points with the same phase, such as between crests, or troughs, or corresponding zero crossings as shown.

Assuming a sinusoidal wave moving at a fixed wave speed, wavelength is inversely proportional to frequency of the wave: waves with higher frequencies have shorter wavelengths, and lower frequencies have longer wavelengths.

Wavelength depends on the medium (for example, vacuum, air, or water) that a wave travels through.

Examples of wave-like phenomena are sound waves, light, and water waves. A sound wave is a variation in air pressure, while in light and other electromagnetic radiation the strength of the electric and the magnetic field vary. Water waves are variations in the height of a body of water. In a crystal lattice vibration, atomic positions vary.

Wavelength is a measure of the distance between repetitions of a shape feature such as peaks, valleys, or zero-crossings, not a measure of how far any given particle moves. For example, in sinusoidal waves over deep water a particle near the water's surface moves in a circle of the same diameter as the wave height, unrelated to wavelength. The range of wavelengths or frequencies for wave phenomena is called a spectrum. The name originated with the visible light spectrum but now can be applied to the entire electromagnetic spectrum as well as to a sound spectrum or vibration spectrum.

Sinusoidal Waves

In linear media, any wave pattern can be described in terms of the independent propagation of sinusoidal components. The wavelength λ of a sinusoidal waveform traveling at constant speed v is given by

$$\lambda = \frac{v}{f},$$

where v is called the phase speed (magnitude of the phase velocity) of the wave and f is the wave's frequency. In a *dispersive* medium, the phase speed itself depends upon the frequency of the wave, making the relationship between wavelength and frequency nonlinear.

In the case of electromagnetic radiation—such as light—in free space, the phase speed is the speed of light, about 3×10^8 m/s. Thus the wavelength of a 100 MHz electromagnetic (radio) wave is about: 3×10^8 m/s divided by 10^8 Hz = 3 metres. The wavelength of visible light ranges from deep red, roughly 700 nm, to violet, roughly 400 nm (for other examples).

For sound waves in air, the speed of sound is 343 m/s (at room temperature and atmospheric pressure). The wavelengths of sound frequencies audible to the human ear (20 Hz–20 kHz) are thus between approximately 17 m and 17 mm, respectively. Note that the wavelengths in audible sound are much longer than those in visible light.

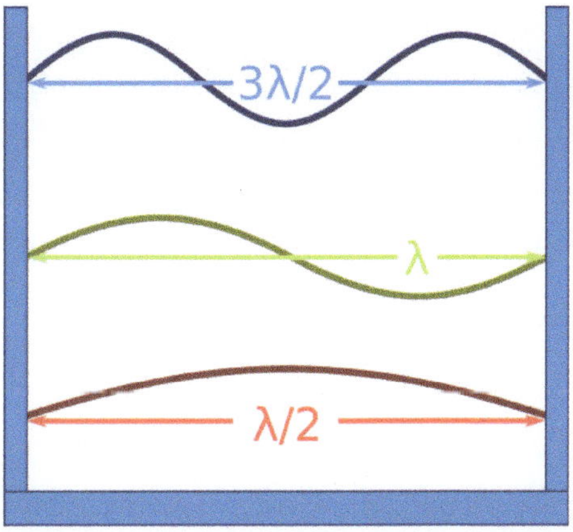

Sinusoidal standing waves in a box that constrains the end points to be nodes will have an integer number of half wavelengths fitting in the box.

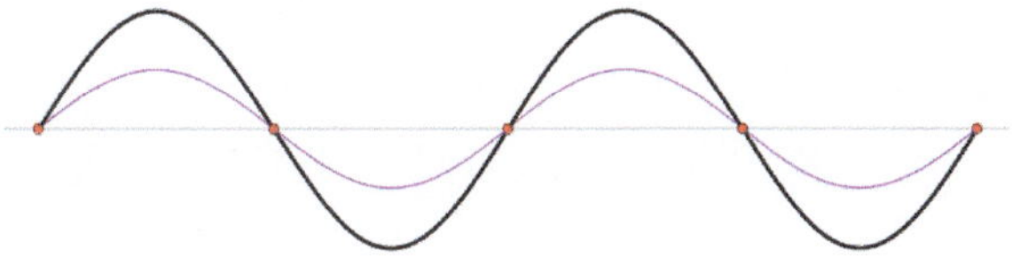

A standing wave (black) depicted as the sum of two propagating waves traveling in opposite directions (red and blue)

Standing Waves

A standing wave is an undulatory motion that stays in one place. A sinusoidal standing wave includes stationary points of no motion, called nodes, and the wavelength is twice the distance between nodes.

The upper figure shows three standing waves in a box. The walls of the box are considered to require the wave to have nodes at the walls of the box (an example of boundary conditions) determining which wavelengths are allowed. For example, for an electromagnetic wave, if the box has ideal metal walls, the condition for nodes at the walls results because the metal walls cannot support a tangential electric field, forcing the wave to have zero amplitude at the wall.

The stationary wave can be viewed as the sum of two traveling sinusoidal waves of oppositely directed velocities. Consequently, wavelength, period, and wave velocity are related just as for a traveling wave. For example, the speed of light can be determined from observation of standing waves in a metal box containing an ideal vacuum.

Mathematical Representation

Traveling sinusoidal waves are often represented mathematically in terms of their velocity v (in the x direction), frequency f and wavelength λ as:

$$y(x,t) = A\cos\left(2\pi\left(\frac{x}{\lambda} - ft\right)\right) = A\cos\left(\frac{2\pi}{\lambda}(x - vt)\right)$$

where y is the value of the wave at any position x and time t, and A is the amplitude of the wave. They are also commonly expressed in terms of wavenumber k (2π times the reciprocal of wavelength) and angular frequency ω (2π times the frequency) as:

$$y(x,t) = A\cos(kx - \omega t) = A\cos(k(x - vt))$$

in which wavelength and wavenumber are related to velocity and frequency as:

$$k = \frac{2\pi}{\lambda} = \frac{2\pi f}{v} = \frac{\omega}{v},$$

or

$$\lambda = \frac{2\pi}{k} = \frac{2\pi v}{\omega} = \frac{v}{f}.$$

In the second form given above, the phase ($kx - \omega t$) is often generalized to ($\mathbf{k}\cdot\mathbf{r} - \omega t$), by replacing the wavenumber k with a wave vector that specifies the direction and wavenumber of a plane wave in 3-space, parameterized by position vector $\mathbf{r}$. In that case, the wavenumber k, the magnitude of $\mathbf{k}$, is still in the same relationship with wavelength as shown above, with v being interpreted as scalar speed in the direction of the wave vector. The first form, using reciprocal wavelength in the phase, does not generalize as easily to a wave in an arbitrary direction.

Generalizations to sinusoids of other phases, and to complex exponentials, are also common. The typical convention of using the cosine phase instead of the sine phase when describing a wave is based on the fact that the cosine is the real part of the complex exponential in the wave

$$Ae^{i(kx - \omega t)}.$$

General Media

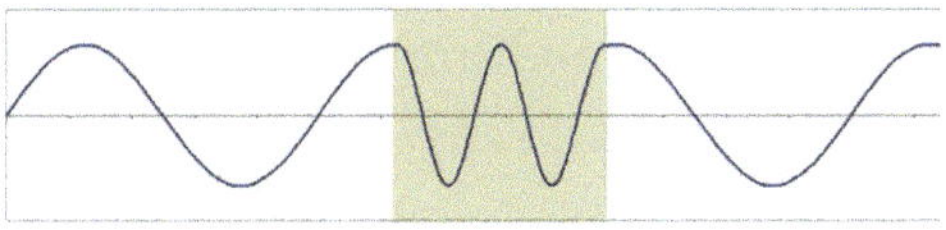

Wavelength is decreased in a medium with slower propagation.

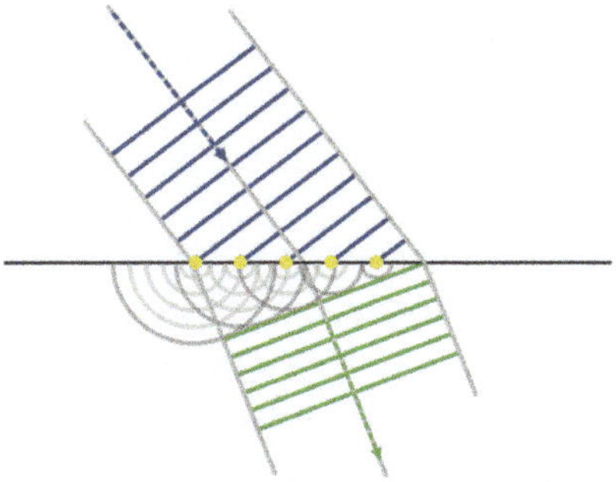

Refraction: upon entering a medium where its speed is lower, the wave changes direction.

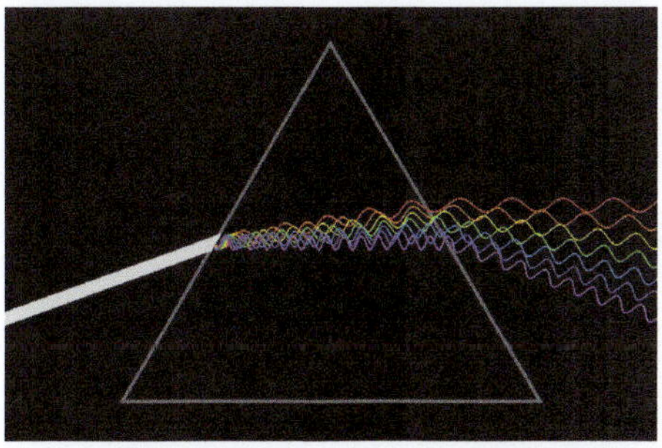

Separation of colors by a prism

The speed of a wave depends upon the medium in which it propagates. In particular, the speed of light in a medium is less than in vacuum, which means that the same frequency will correspond to a shorter wavelength in the medium than in vacuum, as shown in the figure at right.

This change in speed upon entering a medium causes refraction, or a change in direction of waves that encounter the interface between media at an angle. For electromagnetic waves, this change in the angle of propagation is governed by Snell's law.

The wave velocity in one medium not only may differ from that in another, but the velocity typically varies with wavelength. As a result, the change in direction upon entering a different medium changes with the wavelength of the wave.

For electromagnetic waves the speed in a medium is governed by its *refractive index* according to

$$v = \frac{c}{n(\lambda_0)},$$

where c is the speed of light in vacuum and $n(\lambda_0)$ is the refractive index of the medium at wavelength λ_0, where the latter is measured in vacuum rather than in the medium. The corresponding wavelength in the medium is

$$\lambda = \frac{\lambda_0}{n(\lambda_0)}.$$

When wavelengths of electromagnetic radiation are quoted, the wavelength in vacuum usually is intended unless the wavelength is specifically identified as the wavelength in some other medium. In acoustics, where a medium is essential for the waves to exist, the wavelength value is given for a specified medium.

The variation in speed of light with vacuum wavelength is known as dispersion, and is also responsible for the familiar phenomenon in which light is separated into component colors by a prism. Separation occurs when the refractive index inside the prism varies with wavelength, so different wavelengths propagate at different speeds inside the prism, causing them to refract at different angles. The mathematical relationship that describes how the speed of light within a medium varies with wavelength is known as a dispersion relation.

Nonuniform Media

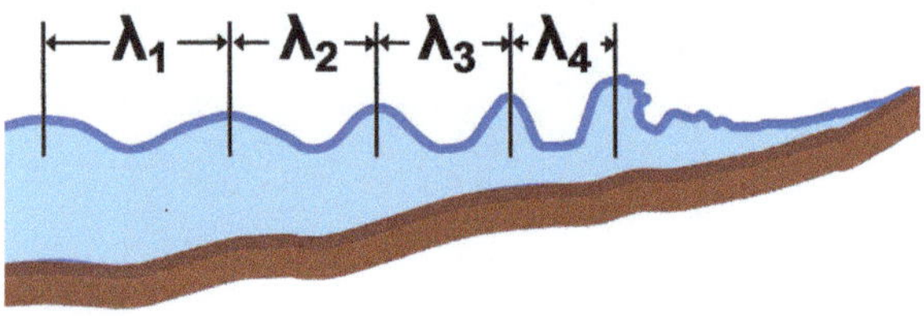

Various local wavelengths on a crest-to-crest basis in an ocean wave approaching shore

Wavelength can be a useful concept even if the wave is not periodic in space. For example, in an ocean wave approaching shore, shown in the figure, the incoming wave undulates with a varying *local* wavelength that depends in part on the depth of the sea floor compared to the wave height. The analysis of the wave can be based upon comparison of the local wavelength with the local water depth.

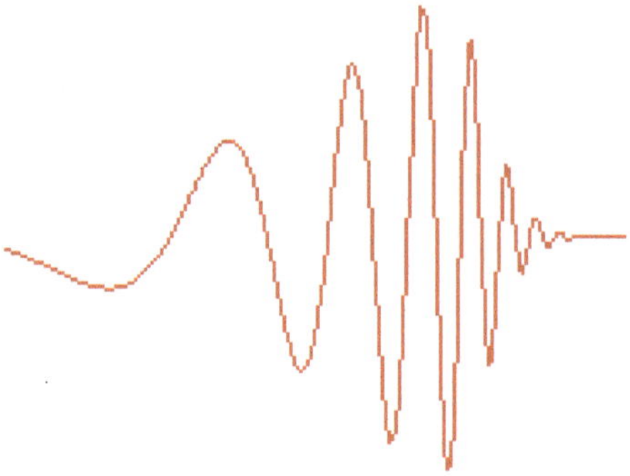

A sinusoidal wave travelling in a nonuniform medium, with loss

Waves that are sinusoidal in time but propagate through a medium whose properties vary with position (an *inhomogeneous* medium) may propagate at a velocity that varies with position, and as a result may not be sinusoidal in space. The figure at right shows an example. As the wave slows down, the wavelength gets shorter and the amplitude increases; after a place of maximum response, the short wavelength is associated with a high loss and the wave dies out.

The analysis of differential equations of such systems is often done approximately, using the *WKB method* (also known as the *Liouville–Green method*). The method integrates phase through space using a local wavenumber, which can be interpreted as indicating a "local wavelength" of the solution as a function of time and space. This method treats the system locally as if it were uniform with the local properties; in particular, the local wave velocity associated with a frequency is the only thing needed to estimate the corresponding local wavenumber or wavelength. In addition, the method computes a slowly changing amplitude to satisfy other constraints of the equations or of the physical system, such as for conservation of energy in the wave.

Crystals

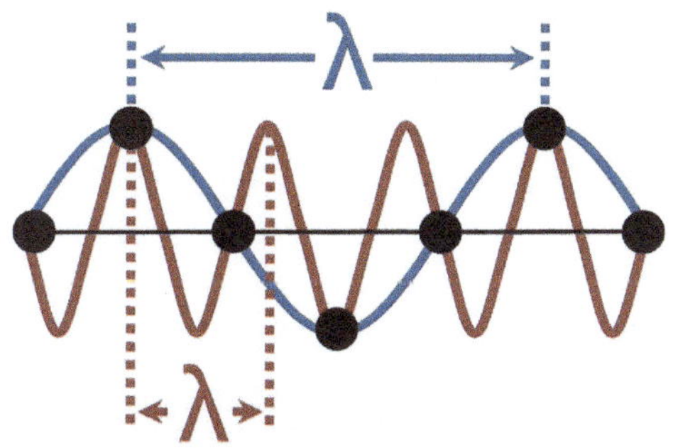

A wave on a line of atoms can be interpreted according to a variety of wavelengths.

Waves in crystalline solids are not continuous, because they are composed of vibrations of discrete particles arranged in a regular lattice. This produces aliasing because the same vibration can be considered to have a variety of different wavelengths, as shown in the figure. Descriptions using more than one of these wavelengths are redundant; it is conventional to choose the longest wavelength that fits the phenomenon. The range of wavelengths sufficient to provide a description of all possible waves in a crystalline medium corresponds to the wave vectors confined to the Brillouin zone.

This indeterminacy in wavelength in solids is important in the analysis of wave phenomena such as energy bands and lattice vibrations. It is mathematically equivalent to the aliasing of a signal that is sampled at discrete intervals.

More General Waveforms

The concept of wavelength is most often applied to sinusoidal, or nearly sinusoidal, waves, because in a linear system the sinusoid is the unique shape that propagates with no shape change – just a phase change and potentially an amplitude change. The wavelength (or alternatively wavenumber or wave vector) is a characterization of the wave in space, that is functionally related to its frequency, as constrained by the physics of the system. Sinusoids are the simplest traveling wave solutions, and more complex solutions can be built up by superposition.

Near-periodic waves over shallow water

In the special case of dispersion-free and uniform media, waves other than sinusoids propagate with unchanging shape and constant velocity. In certain circumstances, waves of unchanging shape also can occur in nonlinear media; for example, the figure shows ocean waves in shallow water that have sharper crests and flatter troughs than those of a sinusoid, typical of a cnoidal wave, a traveling wave so named because it is described by the Jacobi elliptic function of m-th order, usually denoted as $cn(x; m)$. Large-amplitude ocean waves with certain shapes can propagate unchanged, because of properties of the nonlinear surface-wave medium.

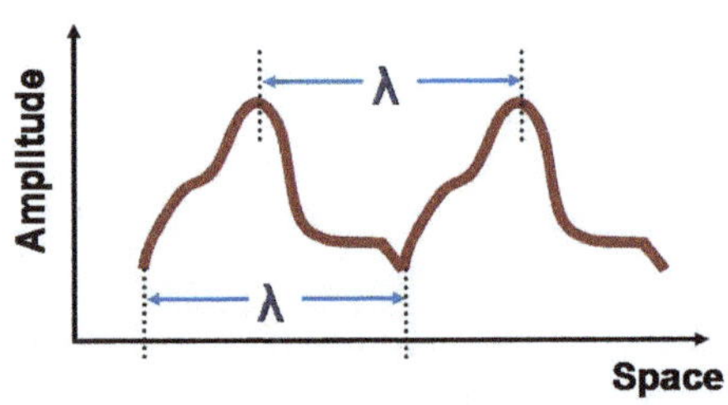

Wavelength of a periodic but non-sinusoidal waveform.

If a traveling wave has a fixed shape that repeats in space or in time, it is a *periodic wave*. Such waves are sometimes regarded as having a wavelength even though they are not sinusoidal. As shown in the figure, wavelength is measured between consecutive corresponding points on the waveform.

Wave Packets

Localized wave packets, "bursts" of wave action where each wave packet travels as a unit, find application in many fields of physics. A wave packet has an *envelope* that describes the overall amplitude of the wave; within the envelope, the distance between adjacent peaks or troughs is sometimes called a *local wavelength*. An example is shown in the figure. In general, the *envelope* of the wave packet moves at a different speed than the constituent waves.

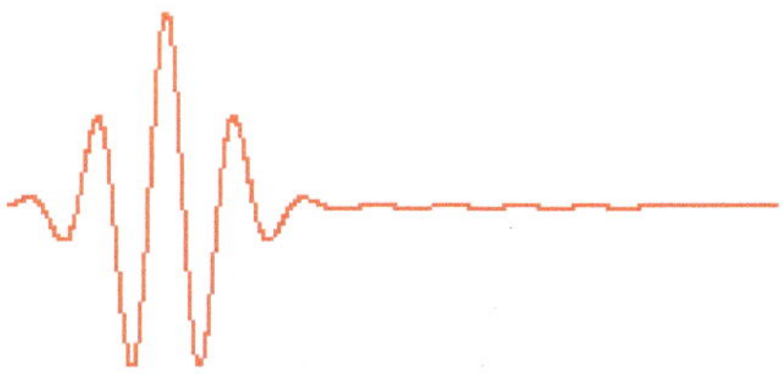

A propagating wave packet

Using Fourier analysis, wave packets can be analyzed into infinite sums (or integrals) of sinusoidal waves of different wavenumbers or wavelengths.

Louis de Broglie postulated that all particles with a specific value of momentum p have a wavelength $\lambda = h/p$, where h is Planck's constant. This hypothesis was at the basis of quantum mechanics. Nowadays, this wavelength is called the de Broglie wavelength. For example, the electrons in a CRT display have a De Broglie wavelength of about 10^{-13} m. To prevent the wave function for such a particle being spread over all space, de Broglie proposed using wave packets to represent particles that are localized in space. The spatial spread of the wave packet, and the spread of the wavenumbers of sinusoids that make up the packet, correspond to the uncertainties in the particle's position and momentum, the product of which is bounded by Heisenberg uncertainty principle.

Interference and Diffraction

Double-slit Interference

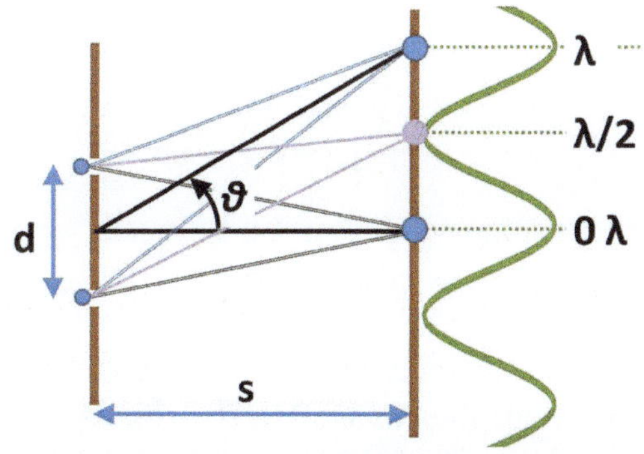

Pattern of light intensity on a screen for light passing through two slits. The labels on the right refer to the difference of the path lengths from the two slits, which are idealized here as point sources.

When sinusoidal waveforms add, they may reinforce each other (constructive interference) or cancel each other (destructive interference) depending upon their relative phase. This phenomenon is used in the interferometer. A simple example is an experiment due to Young where light is passed through two slits. As shown in the figure, light is passed through two slits and shines on a screen. The path of the light to a position on the screen is different for the two slits, and depends upon the angle θ the path makes with the screen. If we suppose the screen is far enough from the slits (that is, *s* is large compared to the slit separation *d*) then the paths are nearly parallel, and the path difference is simply $d \sin \theta$. Accordingly, the condition for constructive interference is:

$$d \sin \theta = m\lambda ,$$

where *m* is an integer, and for destructive interference is:

$$d \sin \theta = (m + 1/2)\lambda .$$

Thus, if the wavelength of the light is known, the slit separation can be determined from the interference pattern or *fringes*, and *vice versa*.

For multiple slits, the pattern is

$$I_q = I_1 \sin^2\left(\frac{q\pi g \sin \alpha}{\lambda}\right) / \sin^2\left(\frac{\pi g \sin \alpha}{\lambda}\right),$$

where *q* is the number of slits, and *g* is the grating constant. The first factor, I_1, is the single-slit result, which modulates the more rapidly varying second factor that depends upon the number of slits and their spacing. In the figure I_1 has been set to unity, a very rough approximation.

It should be noted that the effect of interference is to *redistribute* the light, so the energy contained in the light is not altered, just where it shows up.

Single-slit Diffraction

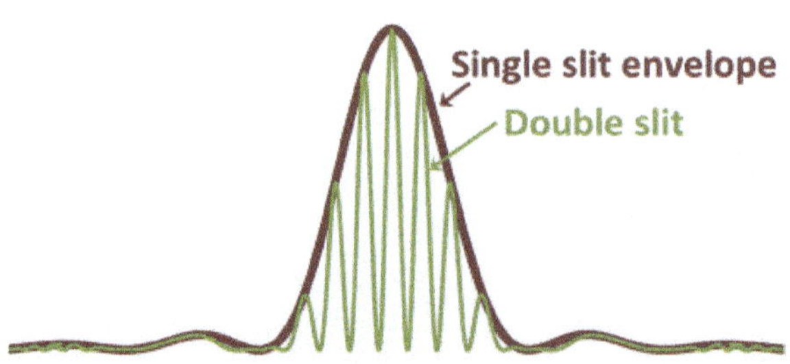

Diffraction pattern of a double slit has a single-slit envelope.

The notion of path difference and constructive or destructive interference used above for the double-slit experiment applies as well to the display of a single slit of light intercepted on a screen. The main result of this interference is to spread out the light from the narrow slit into a broader image on the screen. This distribution of wave energy is called diffraction.

Two types of diffraction are distinguished, depending upon the separation between the source and the screen: Fraunhofer diffraction or far-field diffraction at large separations and Fresnel diffraction or near-field diffraction at close separations.

In the analysis of the single slit, the non-zero width of the slit is taken into account, and each point in the aperture is taken as the source of one contribution to the beam of light (*Huygen's wavelets*). On the screen, the light arriving from each position within the slit has a different path length, albeit possibly a very small difference. Consequently, interference occurs.

In the Fraunhofer diffraction pattern sufficiently far from a single slit, within a small-angle approximation, the intensity spread *S* is related to position *x* via a squared sinc function:

$$S(u) = \text{sinc}^2(u) = \left(\frac{\sin \pi u}{\pi u}\right)^2 ; \text{ with } \quad u = \frac{xL}{\lambda R},$$

where *L* is the slit width, *R* is the distance of the pattern (on the screen) from the slit, and λ is the wavelength of light used. The function *S* has zeros where *u* is a non-zero integer, where are at *x* values at a separation proportion to wavelength.

Diffraction-limited Resolution

Diffraction is the fundamental limitation on the resolving power of optical instruments, such as telescopes (including radiotelescopes) and microscopes. For a circular aperture, the diffraction-limited image spot is known as an Airy disk; the distance *x* in the single-slit diffraction formula is replaced by radial distance *r* and the sine is replaced by $2J_1$, where J_1 is a first order Bessel function.

The resolvable *spatial* size of objects viewed through a microscope is limited according to the Rayleigh criterion, the radius to the first null of the Airy disk, to a size proportional to the wavelength of the light used, and depending on the numerical aperture:

$$r_{Airy} = 1.22 \frac{\lambda}{2\text{NA}},$$

where the numerical aperture is defined as $\text{NA} = n \sin\theta$ for θ being the half-angle of the cone of rays accepted by the microscope objective.

The *angular* size of the central bright portion (radius to first null of the Airy disk) of the image diffracted by a circular aperture, a measure most commonly used for telescopes and cameras, is:

$$\delta = 1.22 \frac{\lambda}{D},$$

where λ is the wavelength of the waves that are focused for imaging, *D* the entrance pupil diameter of the imaging system, in the same units, and the angular resolution δ is in radians.

As with other diffraction patterns, the pattern scales in proportion to wavelength, so shorter wavelengths can lead to higher resolution.

Subwavelength

The term *subwavelength* is used to describe an object having one or more dimensions smaller than the length of the wave with which the object interacts. For example, the term *subwavelength-diameter optical fibre* means an optical fibre whose diameter is less than the wavelength of light propagating through it.

A subwavelength particle is a particle smaller than the wavelength of light with which it interacts. Subwavelength apertures are holes smaller than the wavelength of light propagating through them. Such structures have applications in extraordinary optical transmission, and zero-mode waveguides, among other areas of photonics.

Subwavelength may also refer to a phenomenon involving subwavelength objects; for example, subwavelength imaging.

Angular Wavelength

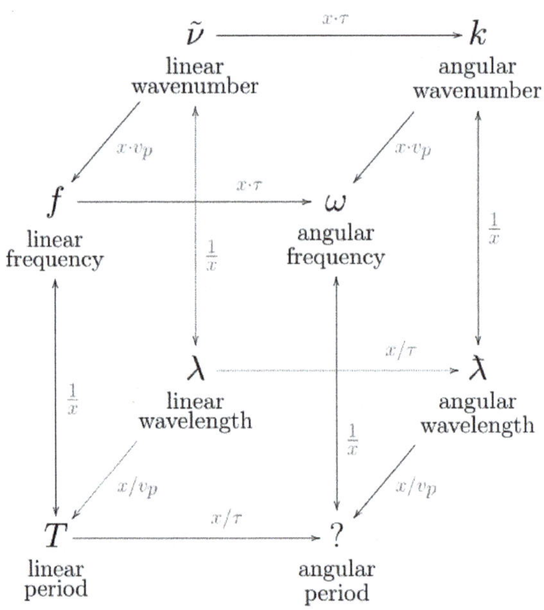

v_p = wave speed (phase velocity)
$\tau = 2\pi$

Relationship between wavelength, angular wavelength, and other wave properties

A quantity related to the wavelength is the angular wavelength (also known as reduced wavelength), usually symbolized by ƛ (lambda-bar). It is equal to the "regular" wavelength "reduced" by a factor of 2π (ƛ = $\lambda/2\pi$). It is usually encountered in quantum mechanics, where it is used in combination with the reduced Planck constant (symbol $\hbar$, h-bar) and the angular frequency (symbol ω) or angular wavenumber (symbol k).

References

- Anssi Klapuri, "Introduction to Music Transcription", in Signal Processing Methods for Music Transcription, edited by Anssi Klapuri and Manuel Davy, 1–20 (New York: Springer, 2006): p. 8. ISBN 978-0-387-30667-4.
- Hartmann, William Morris (1997). Signals, Sound, and Sensation. Springer. pp. 145, 284, 287. ISBN 1-56396-283-7.
- Robert A. Dobie & Susan B. Van Hemel (2005). Hearing Loss: Determining Eligibility for Social Security Benefits. National Academies Press. pp. 50–51. ISBN 978-0-309-09296-8.
- E. Bruce Goldstein (2001). Blackwell Handbook of Perception (4th ed.). Wiley-Blackwell. p. 381. ISBN 978-0-631-20683-5.
- Richard Lyon & Shihab Shamma (1996). "Auditory Representation of Timbre and Pitch". In Harold L. Hawkins & Teresa A. McMullen. Auditory Computation. Springer. pp. 221–23. ISBN 978-0-387-97843-7.
- Birger Kollmeier; Thomas Brand & B. Meyer (2008). "Perception of Speech and Sound". In Jacob Benesty; M. Mohan Sondhi & Yiteng Huang. Springer Handbook of Speech Processing. Springer. p. 65. ISBN 978-3-540-49125-5.
- Winold, Allen (1975). "Rhythm in Twentieth-Century Music" (chapter 3), Aspects of Twentieth-Century Music. Delone and Wittlich (eds.). Englewood Cliffs, New Jersey: Prentice-Hall. ISBN 0-13-049346-5.
- Dixon Ward, W. (1970) "Musical Perception". In Foundations of Modern Auditory Theory vol. 1, edited by Jerry V. Tobias,[page needed]. New York: Academic Press. ISBN 0-12-691901-1.
- Erickson, Robert (1975). Sound Structure in Music. Berkeley and Los Angeles: University of California Press. ISBN 0-520-02376-5.
- Samson, Jim (1977). Music in Transition: A Study of Tonal Expansion and Atonality, 1900-1920. New York: W. W. Norton & Company. ISBN 0-393-02193-9.

7

Audiology: An Overview

Audiology is a branch of science that studies topics such as hearing and the disorders related to hearing. The aspects explained in the following section are audiometers, audiograms, audiometrists, audiometry and pure tone audiometry. This section is an overview of the subject matter incorporating all the main features of audiology.

Audiology

Audiology is a branch of science that studies hearing, balance, and related disorders. Its practitioners, who treat those with hearing loss and proactively prevent related damage are audiologists. Employing various testing strategies (e.g. hearing tests, otoacoustic emission measurements, videonystagmography, and electrophysiologic tests), audiology aims to determine whether someone can hear within the normal range, and if not, which portions of hearing (high, middle, or low frequencies) are affected and to what degree. If an audiologist determines that a hearing loss or vestibular abnormality is present he or she will provide recommendations to a patient as to what options (e.g. hearing aid, cochlear implants, appropriate medical referrals) may be of assistance.

In addition to testing hearing, audiologists can also work with a wide range of clientele in rehabilitation (individuals with tinnitus, auditory processing disorders, cochlear implant users and/or hearing aid users), from pediatric populations to veterans and may perform assessment of tinnitus and the vestibular system.

Audiologist

An audiologist is a health-care professional specializing in identifying, diagnosing, treating and monitoring disorders of the auditory and vestibular system portions of the ear. Audiologists are trained to diagnose, manage and/or treat hearing, tinnitus, or balance problems. They dispense, manage, and rehabilitate hearing aids and assess candidacy for and map cochlear implants. They counsel families through a new diagnosis of hearing loss in infants, and help teach coping and compensation skills to late-deafened adults. They also help design and implement personal and industrial hearing safety programs, newborn hearing screening programs, school hearing screening programs, and provide special fitting ear plugs and other hearing protection devices to help prevent hearing loss. Audiologists are trained to evaluate peripheral vestibular disorders originating from inner ear pathologies. They also provide treatment for certain vestibular and balance disorders such as Benign Paroxysmal Positional Vertigo (BPPV). In addition, many audiologists work as auditory scientists in a research capacity.

Audiologists have training in anatomy and physiology, hearing aids, cochlear implants, electrophysiology, acoustics, psychophysics, neurology, vestibular function and assessment, balance disorders, counseling and sign language. Audiologists also run neonatal hearing screening programme which has been made compulsory in many hospitals in US, UK and India. An Audiologist usually graduates with one of the following qualifications: MSc(Audiology), Au.D., STI, PhD, or ScD, depending the program and country attended.

History

The use of the terms "Audiology" and "Audiologist" in publications has been traced back only as far as 1946. The original creator of the term remains unknown, but Berger identified possible originators as Mayer BA Schier, Willard B Hargrave, Stanley Nowak, Norman Canfield, or Raymond Carhart. In a biographical profile by Robert Galambos, Hallowell Davis is credited with coining the term in the 1940s, saying the then-prevalent term "auricular training" sounded like a method of teaching people how to wiggle their ears.

The first US university course for audiologists was offered by Carhart at Northwestern University, in 1946. Audiology was born of hearing aid dispensers to address the hearing damage from World War II veterans.

Requirements

Australia

In Australia Audiologists must hold a Masters in Audiology, Masters of Clinical Audiology or Masters of Audiology Studies or alternatively bachelor's degree from overseas certified by the VETASSESS. Audiologists in Australia are not required to be a member of any professional body. But to dispense hearing aids to eligible pensioners and eligible war veterans one must have two years clinical experience and be registered with an approved body such as Audiology Australia (AudA) or the Australian College of Audiology (ACAud). In Australia, by general definition an 'Audiologist' - is a University graduate with postgraduate qualifications in Audiology or equivalent training. Audiologists have broad responsibilities and expertise in all non-medical areas of hearing services including complex hearing assessment and rehabilitation of hearing impairment (which includes hearing aid prescription, fitting and management). An 'Audiometrist' - has completed a TAFE Certificate Course in hearing aid audiometry and/or received in-house training from the hearing aid industry.

Audiology Australia via The Code of Ethics and the Practice Standards, governs the professional practice of audiology for members of Audiology Australia. To meet these high standards Members undertake professional development to enable them to maintain appropriate skills and learning in their areas of professional practice. The purpose of the Continuing Professional Development (CPD) program is to provide a structure that enables members to formally document the ways in which they manage and extend their professional skills and knowledge. The Audiology Australia CPD program provides recognition and encouragement for the continuing participation of the members in high quality professional development. The CPD program enable members to demonstrate to external stakeholders (clients, employers, government, the community and other professional groups) their commitment to the highest standards of professional competence for Audiologists. The CPD program is focused on maintaining a high quality practice of audiology and is tied to the Certificate of Clinical Practice (CCP). Audiology Australia members who are not engaged in clinical practice are not required to formally document their CPD and will not be issued with a Certificate of Clinical Practice. ACAud Members are required to demonstrate their professional competence and are assessed against ACAud's Professional Competency Standards. Recognised Competencies are shown on a Member's Certificate of Competency that is prominently displayed in the Member's clinic and renewed annually.

There are six Universities in Australia that offer graduate programs (via course work and/or research degrees) in Audiology for local and overseas students. All programs offered are recognised as eligible for membership of the Society - Audiology Australia and the International Society of Audiology (ISA).

- University of Western Australia
- University of Melbourne
- Flinders University
- La Trobe University
- Macquarie University
- University of Queensland

Canada

In Canada, a Masters of Science (M.Sc.) is the minimum requirement to practice Audiology in the country. The profession is regulated in certain provinces: New Brunswick, Quebec, Ontario, Manitoba, Saskatchewan, Alberta and British Columbia, where it is illegal to practice without being registered as a full member in the appropriate provincial regulatory body.

There are currently five universities in Canada which offer graduate programs in Audiology. Entry requirements typically include specific prerequisite course work in undergraduate studies (usually in phonetics, phonology, acoustics, developmental psychology, perception, anatomy, statistics, physics and research methods) or an additional preparatory year prior to entry into the program and a competitive GPA in the bachelor's degree. The following universities offer the MSc in audiology:

- University of British Columbia (BC)
- Université de Montréal (QC) - French instruction
- University of Western Ontario (ON)
- University of Ottawa (ON) - French instruction
- Dalhousie University (NS)

India

The Government of India established the All India Institute of Speech and Hearing, which has become the country's leading Institute in the field of communication disorders since 1966. The second Audiology & Speech Language Therapy program was started in the same year, at T.N.Medical College and BYL Nair Ch.Hospital in Mumbai. There are currently 20 Universities in India which provide Speech Pathology and Audiology programs. These programs are accredited by Rehabilitation council of India.

To practice audiology, professionals need to have either Bachelors/Master's degree in Audiology and be registered with Rehabilitation Council of India (RCI). Indian speech and hearing association (ISHA) is a professional platform of the audiologist and speech language pathologists where annual conferences are held. The educational institutions awarding degrees need to attain recognition from RCI. Presently there are around 70 institutions which offer at least a graduate degree in the field (BASLP). Approximately 20 colleges offer higher degrees such as MASLP, M.Sc Audiology and Speech Language Pathology. Few such eminent institutions in Northern India are Guru Gobind Singh Medical College located in Faridkot of Punjab, Post Graduate Institute of Medical Education and Research, Chandigarh; Ali Yavar Jung National Institute for the Hearing Handicapped, Mumbai in western India and many more. There are nearly 1000 or more private clinics in India providing speech and hearing services.

An internationally recognized degree, unique multilingual/multicultural background, excellent communication in English has increased the global demand of Indian Audiologists considering the shortage of these graduates especially in western countries. However, in recent times, the profession has been plagued by introduction of short term diploma programs which has resulted in dilution of services particularly in hearing aid dispensing and there have been mass protests by speech and hearing fraternity to stop such courses and curb malpractices.

Malaysia

There are only 3 Malaysian educational institutions offering degrees in Audiology:

- University Kebangsaan Malaysia
- University Science Malaysia
- International Islamic University Malaysia

United Kingdom

There are currently three routes to becoming a Registered Audiologist:

- BSc in Audiology
- MSc in Audiology
- Fast track conversion Diploma for those with a BSc in other relevant science subject, available at Southampton, Manchester, UCL, London and Edinburgh
- BSc(Hons) in Clinical Physiology (Audiology) available at Glasgow Caledonian University. All applicants must be NHS employees.

There are 10 United Kingdom educational institutions offering degrees in Audiology:

- Anglia Ruskin University
- De Montfort University
- University of Manchester
- University of Leeds
- Aston University
- Queen Margaret University
- University of Southampton
- Swansea University
- Middlesex University
- University College London

United States

In the United States, audiologists are regulated by state licensure or registration in all 50 states and the District of Columbia. Starting in 2007, the Doctor of Audiology (Au.D.) became the entry-level degree for clinical practice for some states, with most states expected to follow this requirement very soon, as there are no longer any professional programs in audiology which offer the master's degree. Minimum requirements for the Au.D. degree include a minimum of 75 semester hours of post-baccalaureate study, meeting prescribed competencies, passing a national exam offered by Praxis Series of the Educational Testing Service, and practicum experience that is equivalent to a minimum of 12 months of full-time, supervised experience. Most states have continuing education renewal requirements that must be met to stay licensed. Audiologists can also earn a certificate from the American Speech-Language-Hearing Association or seek board certification through the American Board of Audiology (ABA). Currently there are over 70 Au.D. programs in the United States:

Portugal

The exercise of audiologist profession in Portugal necessarily imply the qualifications Degree in Audiology or legally equivalent as defined in Decree-Law 320/99 of August 11 Article 4.

At present, the degree in Audiology is administered in two Schools:

Health Technology School of Coimbra - Polytechnic Institute of Coimbra

School of Health Sciences - Polytechnic Institute of Porto

In Portugal, research in Audiology is still undeveloped. Most work in this area is mainly on standardization and noise. At this time students of Audiology Colleges may have relevant a role in this area as well as the professionals who attend Masters degree and Doctorates degree.

Audiometer

An audiometer is a machine used for evaluating hearing acuity. They usually consist of an embedded hardware unit connected to a pair of headphones and a test subject feedback button, sometimes controlled by a standard PC. Such systems can also be used with bone vibrators, to test conductive hearing mechanisms.

Audiometers are standard equipment at ENT (ear, nose, throat) clinics and in audiology centers. An alternative to hardware audiometers are software audiometers, which are available in many different configurations. Screening PC-based audiometers use a standard computer. Clinical PC-based audiometers are generally more expensive than software audiometers, but are much more accurate and efficient. They are most commonly used in hospitals, audiology centers and research communities. These audiometers are also used to conduct industrial audiometric testing. Some audiometers even provide a software developer's kit that provides researchers with the capability to create their own diagnostic tests.

Functionality

An audiometer typically transmits recorded sounds such as pure tones or speech to the headphones of the test subject at varying frequencies and intensities, and records the subject's responses to produce an audiogram of threshold sensitivity, or speech understanding profile.

Types

Medical grade audiometers are usually an embedded hardware unit controlled from a PC. Software audiometers which run on a PC are also commercially available, but their accuracy and utility for evaluating hearing loss is questionable due to lack of a calibration standard.

The most common type of audiometer generates pure tones, or transmits parts of speech. Another kind of audiometer is the Bekesy audiometer, in which the subject follows a tone of increasing and decreasing amplitude as the tone is swept through the frequency range by depressing a button when the tone is heard and releasing it when it cannot be heard, crossing back and forth over the threshold of hearing. Bekesy audiometry typically yields lower thresholds and standard deviations than pure tone audiometry.

Standards

Audiometer requirements and the test procedure are specified in IEC 60645, ISO 8253, and ANSI S3.6 standards.

Audiogram

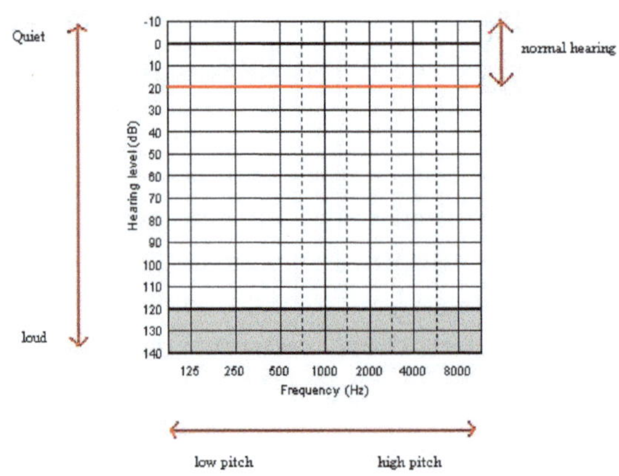

Audiogram

An audiogram is a graph that shows the audible threshold for standardized frequencies as measured by an audiometer. The Y axis represents intensity measured in decibels and the X axis represents frequency measured in Hertz. The threshold of hearing is plotted relative to a standardised curve that represents 'normal' hearing, in dB(HL). They are not the same as equal-loudness contours, which are a set of curves representing equal loudness at different levels, as well as at the threshold of hearing, in absolute terms measured in dB SPL (sound pressure level).

Audiograms are set out with frequency in hertz (Hz) on the horizontal axis, most commonly on a logarithmic scale, and a linear dBHL scale on the vertical axis.

For humans, normal hearing is between −10 dB(HL) and 15 dB(HL), although 0 dB from 250 Hz to 8 kHz is deemed to be 'average' normal hearing.

Hearing thresholds of humans and other mammals can be found by using behavioural hearing tests or physiological tests. An audiogram can be obtained using a behavioural hearing test called Audiometry. For humans the test involves different tones being presented at a specific frequency (pitch) and intensity (loudness). When the person hears the sound they raise their hand or press a button so that the tester knows that they have heard it. The lowest intensity sound they can hear is recorded. The test varies for children, their response to the sound can be a head turn or using a toy. The child learns what they can do when they hear the sound, for example they are taught that when they heard the sound they can put a toy man in a boat. This is referred to as conditioned play audiometry. Visual reinforcement audiometry is also used with children. When the child hears the sound, they look in the direction the sound came from and are reinforced with a light and/or animated toy. A similar technique can be used when testing some animals but instead of a toy, food can be used as a reward for responding to the sound. Physiological tests do not need the patient to respond (Katz 2002). For example when performing the brainstem auditory evoked potentials the patient's brainstem responses are being measured when a sound is played into their ear, or otoacoustic emissions (OAEs) which are generated by a healthy inner ear either spontaneously or evoked by an outside stimulus. In the US, the NIOSH recommends that people who are regularly exposed to hazardous noise have their hearing tested once a year, or every three years otherwise.

Measurement

Audiograms are produced using a piece of test equipment called an audiometer, and this allows different frequencies to be presented to the subject, usually over calibrated headphones, at any specified level. The levels are, however, not absolute, but weighted with frequency relative to a standard graph known as the minimum audibility curve which is intended to represent a 'normal' hearing. This is not the best threshold found for all subjects, under ideal test conditions, which is represented by around 0 Phon or the threshold of hearing on the equal-loudness contours, but is standardised in an

ANSI standard to a level somewhat higher at 1 kHz . There are several definitions of the minimal audibility curve, defined in different international standards, and they differ significantly, giving rise to differences in audiograms according to the audiometer used. The ASA-1951 standard for example used a level of 16.5 dB(SPL) at 1 kHz whereas the later ANSI-1969/ISO-1963 standard uses 6.5 dB(SPL), and it is common to allow a 10 dB correction for the older standard.

Audiograms and Diagnosing Types of Hearing Loss

Most commonly, "conventional" audiometry (utilising audiograms up to 8 kHz) is used to measure hearing status. For research purposes, or early diagnosis of age-related hearing loss, ultra-high frequency audiograms (up to 20 kHz), requiring special audiometer calibration and headphones, can be measured.

Ideally the audiogram would show a straight line, but in practice everyone is slightly different, and small variations are considered normal. Larger variations, especially below the norm, may indicate hearing impairment which occurs to some extent with increasing age, but may be exacerbated by prolonged exposure to fairly high noise levels such as by living close to an airport or busy road, work related exposure to high noise, or brief exposure to very high sound levels such as gunshot or music in either a loud band or clubs and pubs. Hearing impairment may also be the result of certain diseases such as CMV or Ménière's disease and these can be diagnosed from the shape of the audiogram.

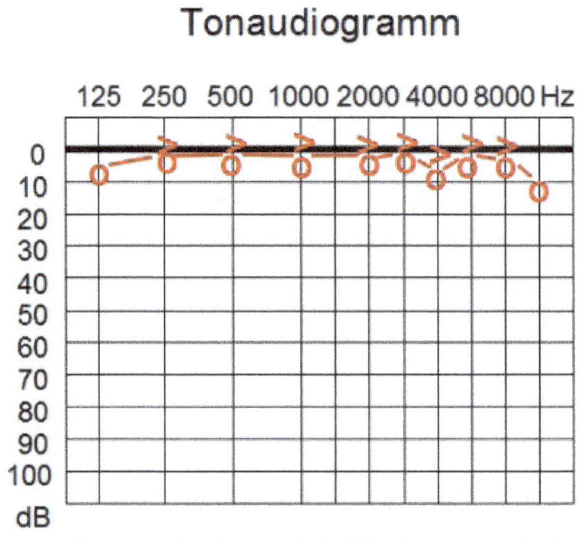

An audiogram showing typical slight hearing variation.

Otosclerosis results in an audiogram with significant loss at all frequencies, often of around 40 dB(HL). A deficiency particularly around 2 kHz (termed a *Carhart notch* in the audiogram) is characteristic of either otosclerosis or a congenital ossicular anomaly.

Ménière's disease results in a severe loss at low frequencies.

Noise induced deafness or sensorineural loss results in loss at high frequencies, especially around 4 kHz and above, depending on the nature of the exposure to loud noise.

Typical examples of audiograms showing conductive, noise induced and age-related hearing loss can be found here.

Audiometrist

An Audiometrist ("to hear"; and from Italian *-metria*, "to measure") or Audiometric Officer, is a health-care professional technician who has received special training in the use of Pure tone audiometry equipment. An audiometrist conducts hearing tests, or "audiometric screening", with an Audiometer to establish hearing levels. The results are represented by an audiogram, and are usually interpreted by an audiologist, or a registered Medical Officer, unless the audiometrist is also an audiologist, with the aim of diagnosing hearing loss.

There are currently some misconceptions regarding the definition of Audiometrist and Audiologist, which vary from country to country. These misconceptions continue to grow, in Australia in particular, which leads to the need for greater communication, less segregation of each other's role within the community and a broader understanding of each other's qualifications.

Audiometrist vs. Audiologist

The word "Audiometrist" is derived from the word Audiometry, as distinct from Audiology. However, the title audiometrist is often used interchangeably with the role of audiologists, thereby making a precise definition of the word more difficult. Audiologists can therefore refer to themselves as audiometrists, although they still perform the tasks of audiologists. The origin of this cross-attribution is unclear, but may stem from the fact that many audiometrists also receive subsequent training in audiology, thereby making them both audiometrists and audiologists. Additionally, preferences for definitions of audiometrists are more varied worldwide, mainly because Audiometry education and licensing requirements vary from country to country, and the use or mis-use of the word in this fashion continues worldwide. A further role confusion arises because many audiometrists are involved directly in the hearing aid industry in an assistive capacity when they get in-house training for hearing aid audiometry, while others apply audiometry purely to Occupational Health. Since audiologists are more qualified, they can do the work of audiometrists, but not the other way round. This may often lead to the assumption that audiometrists are audiologists.

Uk and Ireland

In the UK and Ireland Audiometrists generally differ from Audiologists in that they do not require or hold an Academic Degree or Doctorate in Audiology. Audiometrists are certified under the British and Irish Society of Hearing Aid Audiologists, the Health and Safety Executive (UK and Ireland) and the Health and Safety Authority (Ireland), and the time period of their training is generally shorter than audiologists. Unlike Audiologists, they are not licensed to treat hearing loss with the use of hearing aids.

Australia and New Zealand

As in the UK and Ireland, in Australia and New Zealand Audiometrists can receive certified training in audiometric screening. However, unlike the UK and Ireland, Audiometrists in Australia can prescribe hearing aids to treat a hearing loss if they complete the appropriate training.

An Audiometrist can choose to complete either a Certificate IV in Audiometric Assessment (available only as part of an apprenticeship) or a Diploma of Hearing Device Prescription and Evaluation. Only Audiometrists who complete the Diploma, and the further training below, are qualified to prescribe and fit hearing aids.

Upon graduation of their Diploma, Audiometrists may choose to complete two years supervised clinical experience where they are closely supervised and monitored in a wide variety of techniques. These range from paediatric assessment of clients aged from 3 years to assessment of elderly clients and cover all aspects of screening, diagnostic and prescriptive methods. After this time they are permitted to sit a stringent theoretical and practical exam before a panel of industry experts to be declared competent to prescribe and fit hearing aids. This is a requirement for an Audiometrist to be a member of a professional body.

An Audiologist, who has completed a degree called "Master of Audiology", is required to complete twelve months of supervised clinical experience and be declared competent to be a member of a professional body.

In Australia, the professional bodies for Audiometrists are:

- Australian College of Audiology (ACAud) – Audiologists and Audiometrists; and
- Hearing Aid Audiometrists Society of Australia (HAASA) – Audiometrists only.

Audiometrists providing services under the Australian Government Hearing Services Program must have appropriate membership of a professional body. This also applies to Audiologists.

While an Audiometrist can provide many of the same services as an Audiologist under the Australian Government Hearing Services Program, there are some diagnostic services that can only be provided by Audiologists. Other than referral to an Audiologist or ENT for these services, an Audiometrist does not require approval or supervision by an Audiologist. This does not apply to an Audiometrist with a Certificate IV qualification, who cannot provide unsupervised services under the Australian Government Hearing Services Program.

U.S. and Canada

In the United States and Canada, Audiometrists, sometimes referred to as "Audiometric Technicians", also receive accreditation in audiometric screening, under the Occupational Safety and Health Administration (OSHA)and Master of Healthcare Administration (MHA). When audiometry is applied to industry, audiometrists are sometimes also called Occupational Hearing Conservationists. When Audiometrists become qualified as Audiologists, they are more commonly referred to as Hearing Instrument Specialists, and are licensed hearing professionals that perform diagnostic hearing evaluations, as well as prescribe and fit hearing aids. A Hearing Instrument Specialist in the US or Canada is the equivalent of a Hearing Aid Dispenser in the UK or Ireland.

Audiometry

Audiometry is a branch of Audiology and the science of measuring hearing acuity for variations in sound intensity and pitch and for tonal purity, involving thresholds and differing frequencies. Typically, audiometric tests determine a subject's hearing levels with the help of an audiometer, but may also measure ability to discriminate between different sound intensities, recognize pitch, or distinguish speech from background noise. Acoustic reflex and otoacoustic emissions may also be measured. Results of audiometric tests are used to diagnose hearing loss or diseases of the ear, and often make use of an audiogram.

History and Development

The basic requirements of the field were to be able to produce a repeating sound, some way to attenuate the amplitude, a way to transmit the sound to the subject, and a means to record and interpret the subject's responses to the test.

Mechanical "acuity meters" and Tuning Forks

For many years there were a desultory use of various devices capable of producing sounds of controlled intensity. The first types were clock-like, giving off air-borne sound to the tubes of a stethoscope; the sound distributor head had a valve which could be gradually closed. Another model used a tripped hammer to strike a metal rod and produce the testing sound; in another a tuning fork was struck. The first such measurement device for testing hearing was described by Wolke (1802)

Pure Tone Audiometry and Audiograms

Following development of the induction coil in 1849 and audio transducers (telephone) in 1876, a variety of audiometers were invented in United States and overseas. These early audiometers were known as induction-coil audiometers due to...

- Hughes 1879
- Hartmann 1878

In 1885, Arthur Hartmann designed an "Auditory Chart" which included left and right ear tuning fork representation on the abscissa and percent of hearing along the ordinate.

In 1899, Carl E. Seashore Prof. of Psychology at U. Iowa, United States, introduced the audiometer as an instrument to measure the "keenness of hearing" whether in the laboratory, schoolroom, or office of the psychologist or aurist. The instrument operated on a battery and presented a tone or a click; it had an attenuator set in a scale of 40 steps. His machine became the basis of the audiometers later manufactured at Western Electric.

- Cordia C. Bunch 1919

The concept of a frequency versus sensitivity (amplitude) audiogram plot of human hearing sensitivity was conceived by German physicist Max Wien in 1903. The first vacuum tube implementations, November 1919, two groups of researchers — K.L. Schaefer and G. Gruschke, B. Griessmann and H. Schwarzkopf — demonstrated before the Berlin Oto-logical Society two instruments designed to test hearing acuity. Both were built with vacuum tubes. Their designs were characteristic of the two basic types of electronic circuits used in most electronic audio devices for the next two decades. Neither of the two devices was developed commercially for some time, although the second was to be manufactured under the name "Otaudion." The Western Electric 1A, developed by <who> was built in 1922 in the United States. It was not until 1922 that otolaryngologist Dr. Edmund P. Fowler, and physicists Dr. Harvey Fletcher and Robert Wegel of Western Electric Co. first employed frequency at octave intervals plotted along the abscissa and intensity downward along the ordinate as a degree of hearing loss. Fletcher et al. also coined the term "audiogram" at that time.

With further technologic advances, bone conduction testing capabilities became a standard component of all Western Electric audiometers by 1928.

Electrophysiologic Audiometry

In 1967, Sohmer and Feinmesser were the first to publish ABRs recorded with surface electrodes in humans which showed that cochlear potentials could be obtained non-invasively.

Otoacoustic Audiometry

In 1978, David Kemp reported that sound energy produced by the ear could be detected in the ear canal. The first commercial system for detecting and measuring OAEs was produced in 1988.

Components

The auditory system is composed of epithelial, osseous, vascular, neural and neocortical tissues. The anatomical divisions are external ear canal and tympanic membrane, middle ear, inner ear, VIII auditory nerve, and central auditory processing portions of the neocortex.

The Process of Hearing

Sound waves enter the outer ear and travel through the external auditory canal until they reach the tympanic membrane, causing the membrane and the attached chain of auditory ossicles to vibrate. The motion of the stapes against the oval window sets up waves in the fluids of the cochlea, causing the basilar membrane to vibrate. This stimulates the sensory cells of the organ of Corti, atop the basilar membrane, to send nerve impulses to the central auditory processing areas of the brain, the auditory cortex, where sound is perceived and interpreted.

Types of Audiometry

Subjective Audiometry

Subjective audiometry requires the cooperation of the subject, and relies upon subjective responses which may both qualitative and quantitative, and involve attention (focus), reaction time, etc.

- Differential testing is conducted with a low frequency (usually 512hz) tuning fork. They are used to assess asymmetrical hearing and air/bone conduction differences. They are simple manual physical tests and do not result in an audiogram.
    - Weber test
    - Bing test

- Rinne test
- Schwabach test, a variant of the Rinne test

- Pure tone audiometry is a standardized hearing test in which air conduction hearing thresholds in decibels (db) for a set of fixed frequencies between 250hz and 8,000hz are plotted on an audiogram for each ear independently. A separate set of measurements is made for bone conduction. There is also high frequency Pure Tone Audiometry covering the frequency range above 8000hz to 16,000hz.
- Threshold Equalizing Noise (TEN) test
- Masking Level Difference (MLD) test
- Psychoacoustic (or Psychophysical) Tuning Curve test
- Speech audiometry is a diagnostic hearing test designed to test word or speech recognition. It has become a fundamental tool in hearing-loss assessment. In conjunction with pure-tone audiometry, it can aid in determining the degree and type of hearing loss. Speech audiometry also provides information regarding discomfort or tolerance to speech stimuli and information on word recognition abilities. In addition, information gained by speech audiometry can help determine proper gain and maximum output of hearing aids and other amplifying devices for patients with significant hearing losses and help assess how well they hear in noise. Speech audiometry also facilitates audiological rehabilitation management.

Speech audiometry may include:

- Speech Awareness Threshold
- Speech Recognition Threshold
- Suprathreshold word-recognition
- Sentence testing
- Dichotic listening test
- Loudness levels determination

- Békésy audiometry, also called decay audiometry - audiometry in which the subject controls increases and decreases in intensity as the frequency of the stimulus is gradually changed so that the subject traces back and forth across the threshold of hearing over the frequency range of the test. The test is quick and reliable, so was frequently used in military and industrial contexts.
- Audiometry of children

    - Conditioned play audiometry
    - Behavioral observation audiometry
    - Visual reinforcement audiometry

Objective Audiometry

Objective audiometry is based on physical, acoustic or electrophysiologic measurements and does not depend on the cooperation or subjective responses of the subject.

- Caloric stimulation/reflex test uses temperature difference between hot and cold water or air delivered into the ear to test for neural damage. Caloric stimulation of the ear results in rapid side-to-side eye movements called nystagmus. Absence of nystagmus may indicate auditory nerve damage. This test will often be done as part of another test called electronystagmography.
- Electronystagmography (ENG) uses skin electrodes and an electronic recording device to measure nystagmus evoked by procedures such as caloric stimulation of the ear
- Acoustic immittance audiometry - Immittance audiometry is an objective technique which evaluates middle ear function by three procedures: static immittance, tympanometry, and the measurement of acoustic reflex threshold sensitivity. Immittance audiometry is superior to pure tone audiometry in detecting middle ear pathology.
    - Tympanometry
    - Acoustic Reflex Thresholds (ART)
    - Acoustic reflectometry
    - wide-band absorbance audiometry also called 3D tympanometry
- Evoked potential audiometry
    - N1-P2 Cortical Audio Evoked Potential (CAEP) audiometry
    - Auditory brainstem response (ABR) is a neurologic tests of auditory brainstem function in response to auditory (click) stimuli.
    - Electrocochleography a variant of ABR, tests the impulse transmission function of the cochlea in response to auditory (click) stimuli. It is most often used to detect endolymphatic hydrops in the diagnois/assessment of Meniere’s disease.
    - Audio steady state response (ASSR) audiometry

  - Vestibular evoked myogenic potential (VEMP) test, a variant of ABR that tests the integrity of the saccule
- Otoacoustic emission audiometry - this test can differentiate between the sensory and neural components of sensorineural hearing loss.
  - Distortion Product OtoAcoustic Emissons DPOAE audiometry
  - Transient Evoked OtoAcoustic Emissions TEOAE audiometry
  - Sustained Frequency Otoacoustic emissions (SFOAE) audionetry - At present, SFOAEs are not used clinically.
- In situ audiometry: a technique for measuring not only the affliction of the person's auditory system, but also the characteristics of sound reproduction devices, in-the-canal hearing aids, vents and sound tubes of hearing aids.

Audiograms

The result of most audiometry is an audiogram plotting some measured dimension of hearing, either graphically or tabularly.

The most common type of audiogram is the result of a pure tone audiometry hearing test which plots frequency versus amplitude sensitivity thresholds for each ear along with bone conduction thresholds at 8 standard frequencies from 250hz to 8000hz. A PTA hearing test is the gold standard for evaluation of hearing loss/disability. Other types of hearing tests also generate graphs or tables of results that may be loosely called 'audiograms', but the term is universally used to refer to the result of a PTA hearing test.

Hearing Assessment

Apart from testing hearing, part of the function of audiometry is in assessing or evaluating hearing from the test results. The most commonly used assessment of hearing is the determination of the threshold of audibility, i.e. the level of sound required to be just audible. This level can vary for an individual over a range of up to 5 Decibels from day to day and from determination to determination, but it provides an additional and useful tool in monitoring the potential ill effects of exposure to noise. Before carrying out a hearing test, it is important to obtain information about the person's past medical history, not only concerning the ears but also other conditions which may have a bearing on possible hearing loss detected by an audiometric test. Hearing loss may be unilateral or bilateral, and bilateral hearing loss may not be symmetrical. The most common types of hearing loss, due to age and noise exposure, are usually bilateral and symmetrical. Wax in the ear can also cause hearing loss, so the ear should be examined to see if syringing is needed; also to determine if the eardrum has suffered any damage which may reduce the ability of sound to be transmitted to the middle ear.

Hearing Loss Classification

The primary focus of audiometry is assessment of hearing status and hearing loss, including extent, type and configuration.

- There are four defined degrees of hearing loss: mild, moderate, severe and profound.
- Hearing loss may be divided into four types: conductive hearing loss, sensorineural hearing loss, central auditory processing disorders, and mixed types.
- Hearing loss may be unilateral or bilateal, of sudden onset or progressive, and temporary or permanent.

Hearing loss may be caused by a number of factors including heredity, congenital conditions, age-related (presbycusis) and acquired factors like noise-induced hearing loss, ototoxic chemicals and drugs, infections, and physical trauma.

Clinical Practice

Audiometric testing may be performed by a general practitioner medical doctor, an otolaryngologist (a specialized MD also called an ENT), a CCC-A (Certificate of Clinical Competence in Audiology) audiologist, a certified school audiometrist (a practitioner analogous to an optometrist who tests eyes), and sometimes other trained practitioners. Practitioners are certified by American Board of Audiology (ABA). Practitioners are licensed by various state boards regulating workplace health & safety, occupational professions, or ...

Schools

Noise-induced Hearing Loss

Workplace and environmental noise is the most prevalent cause of hearing loss in the United States and elsewhere.

Research

- Computer modeling of patterns of hearing deficit
- 3D longitudinal profiles of hearing loss including age axis (presbycusis study)

Pure Tone Audiometry

Pure tone audiometry (PTA) is the key hearing test used to identify hearing threshold levels of an individual, enabling determination of the degree, type and configuration of a hearing loss. Thus, providing the basis for diagnosis and management. PTA is a subjective, behavioural measurement of hearing threshold, as it relies on patient response to pure tone stimuli. Therefore, PTA is used on adults and children old enough to cooperate with the test procedure. As with most clinical tests, calibration of the test environment, the equipment and the stimuli to ISO standards is needed before testing proceeds. PTA only measures audibility thresholds, rather than other aspects of hearing such as sound localization and speech recognition. However, there are benefits of using PTA over other forms of hearing test, such as click auditory brainstem response (ABR). PTA provides ear specific thresholds, and uses frequency specific pure tones to give place specific responses, so that the configuration of a hearing loss can be identified. As PTA uses both air and bone conduction audiometry, the type of loss can also be identified via the air-bone gap. Although PTA has many clinical benefits, it is not perfect at identifying all losses, such as 'dead regions' of the cochlea and neuropathies such as auditory processing disorder (APD). This raises the question of whether or not audiograms accurately predict someone's perceived degree of disability.

PTA Procedural Standards

There are both international and British standards regarding the PTA test protocol. The British Society of Audiology (BSA) is responsible for publishing the recommended procedure for PTA, as well as many other audiological procedures. The British recommended procedure is based on international standards. Although there are some differences, the BSA-recommended procedures are in accordance with BS EN ISO 8253-1, which is the international standard for PTA established by the International Organization for Standardization. The BSA-recommended procedures provide a "best practice" test protocol for professionals to follow, increasing validity and allowing standardisation of results across Britain. The British Society of Audiology. Recommended Procedure: Pure Tone air and bone conduction threshold audiometry with and without masking and determination of uncomfortable loudness levels.

Variations

There are cases where conventional PTA is not an appropriate or effective method of threshold testing. Procedural changes to the conventional test method may be necessary with populations who are unable to cooperate with the test in order to obtain hearing thresholds. Sound field audiometry may be more suitable when patients are unable to wear earphones, as the stimuli are usually presented by loudspeaker. A disadvantage of this method is that although thresholds can be obtained, results are not ear specific. In addition, response to pure tone stimuli may be limited, because in a sound field pure tones create standing waves, which alter sound intensity within the sound field. Therefore, it may be necessary to use other stimuli, such as warble tones in sound field testing. There are variations of conventional audiometry testing that are designed specifically for young children and infants, such as behavioral observation audiometry, visual reinforcement audiometry and play audiometry.

Conventional audiometry tests frequencies between 250 hertz (Hz) and 8 kHz, whereas high frequency audiometry tests in the region of 8 kHz-16 kHz. Some environmental factors, such as ototoxic medication and noise exposure, appear to be more detrimental to high frequency sensitivity than to that of mid or low frequencies. Therefore, high frequency audiometry is an effective method of monitoring losses that are suspected to have been caused by these factors. It is also effective in detecting the auditory sensitivity changes that occur with aging.

Cross hearing and Interaural Attenuation

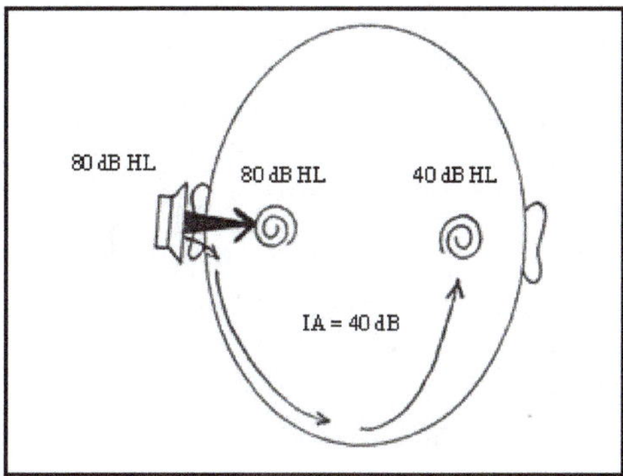

Figure : Interaural attenuation with air conduction.

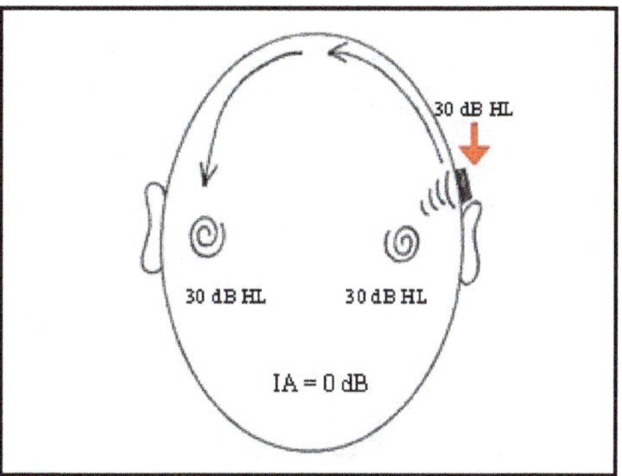

Figure : Interaural attenuation with bone conduction

When sound is applied to one ear the contralateral cochlea can also be stimulated to varying degrees, via vibrations through the bone of the skull. When the stimuli presented to the test ear stimulates the cochlea of the non-test ear, this is known as cross hearing. Whenever it is suspected that cross hearing has occurred it is best to use masking. This is done by temporarily elevating the threshold of the non-test ear, by presenting a masking noise at a predetermined level. This prevents the non-test ear from detecting the test signal presented to the test ear. The threshold of the test ear is measured at the same time as presenting the masking noise to the non-test ear. Thus, thresholds obtained when masking has been applied, provide an accurate representation of the true hearing threshold level of the test ear.

A reduction or loss of energy occurs with cross hearing, which is referred to as interaural attenuation (IA) or transcranial transmission loss. IA varies with transducer type. It varies from 40 dB to 80 dB with supra-aural headphones. However, with insert earphones it is in the region of 55 dB. The use of insert earphones reduces the need for masking, due to the greater IA which occurs when they are used.

Air conduction results in isolation, give little information regarding the type of hearing loss. When the thresholds obtained via air conduction are examined alongside those achieved with bone conduction, the configuration of the hearing loss can be determined. However, with bone conduction (performed by placing a vibrator on the mastoid bone behind the ear), both cochleas are stimulated. IA for bone conduction ranges from 0-20 dB. Therefore, conventional audiometry is ear specific, with regards to both air and bone conduction audiometry, when masking is applied.

PTA Thresholds and Hearing Disability

PTA is described as the gold standard for assessment of a hearing loss. But, how accurate is PTA at classifying the hearing loss of an individual, in terms of hearing impairment and hearing disability? Hearing impairment is defined by the World Health Organisation (WHO) as a hearing loss with thresholds higher than 25db in one or both ears. The degree of hearing loss is classified as mild, moderate, severe or profound. The results of PTA are a good indicator of hearing impairment.

Hearing disability is defined by the WHO as a reduction in ability to hear sounds in quiet and background noise (compared to people with normal hearing), which is caused by a hearing impairment. Several studies have investigated whether self-reported hearing problems (via questionnaires and interviews) were associated with the results from PTA. The findings of these studies indicate that in general, the results of PTA correspond to self-reported hearing problems (i.e. hearing disability). However, for some individuals this is not the case. Therefore, the results of PTA cannot be used to ascertain an individual's hearing disability.

Hearing impairment (based on the audiogram) and auditory handicap (based on speech discrimination in noise) data was reviewed by Reinier Plomp. This led to the formulation of equations, which described the consequences of a hearing loss on speech intelligibility. The results of this review indicated that there were two factors of a hearing loss, which were involved in the effect on speech intelligibility. These factors were named Factor A and Factor D. Factor A affected speech intelligibility by attenuating the speech, whereas Factor D affected speech intelligibility by distorting the speech.

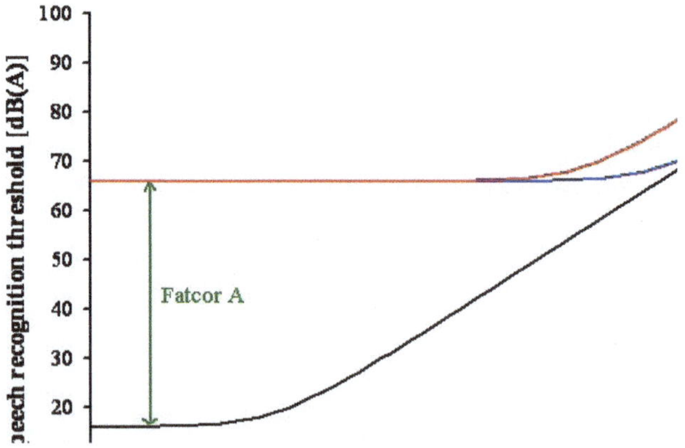

Figure : Speech recognition threshold (SRT) in noise. To aid explanation of this concept the CHL and the SNHL have the same magnitude of hearing loss (50 dBHL). The horizontal part of the curves is where the noise is inaudible. Thus, there is no masking effect on the SRT. The horizontal portion of the curve for the SNHL and CHL extends further than that for a normal hearing person, as the noise needs to become audible to become a problem. Thus, more noise has to be applied, to produce a masking effect. At the right hand side of the graph, to identify 50% of the speech correctly, the speech needs to much more intense than in the quiet. This is because at this end of the graph, the noise is very loud whether the person has a hearing loss or not. There is a transition between these two areas described. Factor A is a problem only in low noise levels, whereas Factor D is a problem when the noise level is high.

Speech recognition threshold (SRT) is defined as the sound pressure level at which 50% of the speech is identified correctly. For a person with a conductive hearing loss (CHL) in quiet, the SRT needs to be higher than for a person with normal hearing. The increase in SRT depends on the degree of hearing loss only, so Factor A reflects the audiogram of that person. In noise, the person with a CHL has the same problem as the person with normal hearing.

For a person with a Sensorineural hearing loss (SNHL) in quiet, the SRT also needs to be higher than for a person with normal hearing. This is because the only factor that is important in quiet for a CHL and a SNHL is the audibility of the sound, which corresponds to Factor A. In noise, the person with a SNHL requires a better signal-to-noise ratio to achieve the same performance level, as the person with normal hearing and the person with a CHL. This shows that in noise, Factor A is not enough to explain the problems of a person with a SNHL. Therefore, there is another problem present, which is Factor D. At present, it is not known what causes Factor D. Thus, in noise the audiogram is irrelevant. It is the type of hearing loss that is important in this situation.

These findings have important implications for the design of hearing aids. As hearing aids at present can compensate for Factor A, but this is not the case for Factor D. This could be why hearing aids are not satisfactory for a lot of people.

Audiograms and Hearing Loss

The shape of the audiogram resulting from PTA audiometry gives an indication of the type of hearing loss as well as possible causes. Conductive hearing loss due to disorders of the middle ear shows as a flat increase in thresholds cross the frequency range. Sensorineural hearing loss will have a contoured shape depending on the cause. Presbycusis or age-related hearing loss for example is characterized by a high frequency roll-off (increase in thresholds). Noise-induced hearing loss has a characteristic notch at 4000Hz. Other coutours may indicate other causes for the hearing loss.

Auditory-verbal Therapy

Auditory-verbal therapy is a method for teaching deaf children to listen and speak using their residual hearing in addition to the constant use of amplification devices such as hearing aids, FM devices, and cochlear implants. Auditory-verbal therapy emphasizes speech and listening.

Auditory verbal therapy enables deaf and hard of hearing children to participate more fully in mainstream school and hearing society. In 2006, Rhoades published an article looking at whether the Auditory Verbal approach was justified. Using Frattali's system for weighing the strength of treatment evidence based on the degree of scientific rigour, she found 7 studies concerning the AV approach of class II level evidence. She concluded that even with the minimal number of studies, there was evidence "for the validity and effectiveness of the A-V intervention model". Following on from Rhoades' work, Hogan et al. (2008; 2010) published outcomes for children in an AV programme which were essentially quasi-experimental studies in that the assessment tool used had been standardised on typically developing children and so a comparison could be drawn between the age equivalence scores obtained by the children on the study and their chronological or 'hearing' age. (The hearing age of a child is the duration of time since the child has had optimal access to sound.) In these studies, the average rate of language development increased significantly during the time on the programme compared to the average rate of language development prior to embarking on the programme. Dornan et al. (2009) reviewed the evidence for Auditory Verbal Therapy and concluded that while it was not possible to demonstrate a cause and effect relationship between the intervention and the outcome, increasing the strength of evidence for outcomes was vital. In 2010, Dornan et al. published a longitudinal study comparing the outcomes of 29 children enrolled on an AV programme with age-matched controls. They found AV to be an effective intervention.

First Voice (a membership body which represents organisations that provide listening and spoken language therapy in Australia and New Zealand) published a study in February 2015 based on information on more than 600 children with hearing loss. When listening and spoken language outcomes of children with a hearing loss were compared to those with typical hearing most children attending AVT had 'scores within or above the average range for typical hearing children for language (74.4%-75.6%), vocabulary (79.6%) and speech performance (71.5%). When children with additional disabilities were removed from the analysis, the number of children within or above the average range for typical hearing children increased, except for speech performance (language performance=77.9%-80.2%; vocabulary performance= 83.1%; speech performance=73.1%).' (First Voice, 2015 pp 5).

In the UK, access to auditory verbal therapy is currently available at two cochlear implant centres (Paediatric Cochlear Implant Programme, Manchester Auditory Implant Programme and Midlands Hearing Implant Programme - Children's Service). Outside of the NHS AVT is available via a national charity called Auditory Verbal^{UK} (AV^{UK})(Oxfordshire and London), The Speech and Hearing and Language Centre, Christopher Place in London and at the Auditory Verbal Lounge in Nottingham.

Auditory System

The auditory system is the sensory system for the sense of hearing. It includes both the sensory organs (the ears) and the auditory parts of the sensory system.

Structure

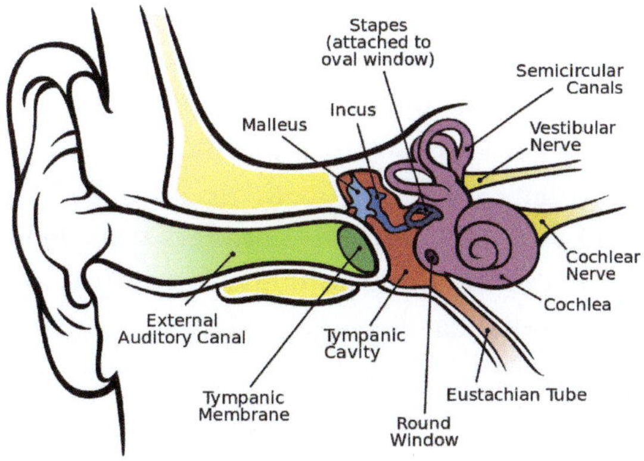

Anatomy of the human ear (The length of the auditory canal is exaggerated in this image.)

Outer Ear

The folds of cartilage surrounding the ear canal are called the pinna. Sound waves are reflected and attenuated when they hit the pinna, and these changes provide additional information that will help the brain determine the direction from which the sounds came.

The sound waves enter the auditory canal, a deceptively simple tube. The ear canal amplifies sounds that are between 3 and 12 kHz. At the far end of the ear canal is the tympanic membrane, which marks the beginning of the middle ear.

Middle Ear

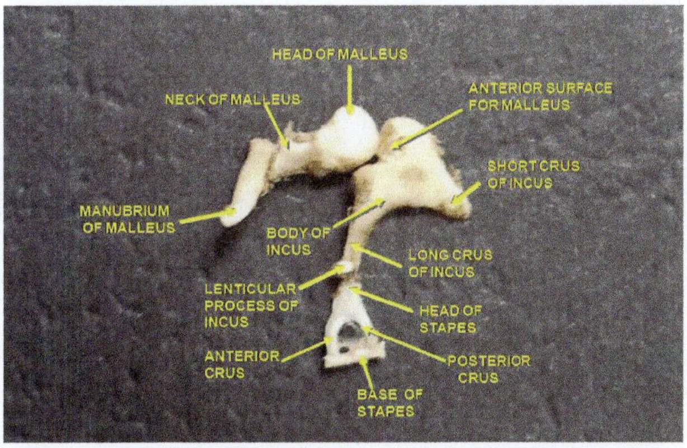

Auditory ossicles from a deep dissection of the tympanic cavity

Sound waves travel through the ear canal and hit the tympanic membrane, or eardrum. This wave information travels across the air-filled middle ear cavity via a series of delicate bones: the malleus (hammer), incus (anvil) and stapes (stirrup). These ossicles act as a lever, converting the lower-pressure eardrum sound vibrations into higher-pressure sound vibrations at another, smaller membrane called the oval window or vestibular window. The manubrium (handle) of the malleus articulates with the tympanic membrane, while the footplate (base) of the stapes articulates with the oval window. Higher pressure is necessary at the oval window than at the typanic membrane because the inner ear beyond the oval window contains liquid rather than air. The stapedius reflex of the middle ear muscles helps protect the inner ear from damage by reducing the transmission of sound energy when the stapedius muscle is activated in response to sound. The middle ear still contains the sound information in wave form; it is converted to nerve impulses in the cochlea.

Inner Ear

The inner ear consists of the cochlea and several non-auditory structures. The cochlea has three fluid-filled sections, and supports a fluid wave driven by pressure across the basilar membrane separating two of the sections. Strikingly, one section, called the cochlear duct or *scala media,* contains endolymph, a fluid similar in composition to the intracellular fluid found inside cells. The organ of Corti is located in this duct on the basilar membrane, and transforms mechanical waves to electric signals in neurons. The other two sections are known as the *scala tympani* and the *scala vestibuli*; these are located within the bony labyrinth, which is filled with fluid called perilymph, similar in composition to cerebrospinal fluid. The chemical difference between the fluids endolymph and perilymph fluids is important for the function of the inner ear due to electrical potential differences between potassium and calcium ions.

The plan view of the human cochlea (typical of all mammalian and most vertebrates) shows where specific frequencies occur along its length. The frequency is an approximately exponential function of the length of the cochlea within the Organ of Corti. In some species, such as bats and dolphins, the relationship is expanded in specific areas to support their active sonar capability.

Organ of Corti

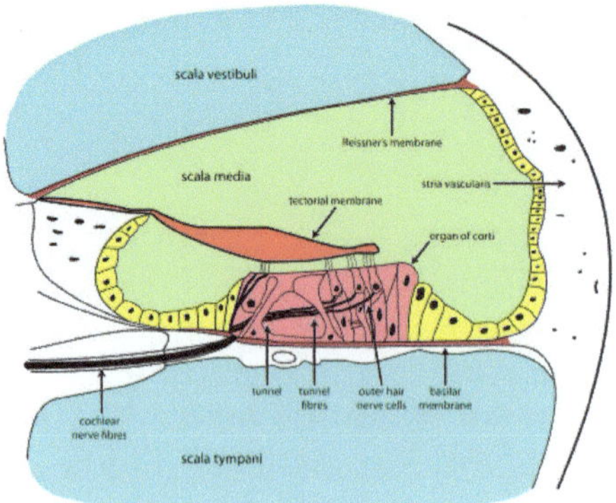

The organ of Corti located at the *scala media*

The organ of Corti forms a ribbon of sensory epithelium which runs lengthwise down the cochlea's entire *scala media*. Its hair cells transform the fluid waves into nerve signals. The journey of countless nerves begins with this first step; from here, further processing leads to a panoply of auditory reactions and sensations.

Hair Cell

Hair cells are columnar cells, each with a bundle of 100-200 specialized cilia at the top, for which they are named. There are two types of hair cells. Inner hair cells are the mechanoreceptors for hearing: they transduce the vibration of sound into electrical activity in nerve fibers, which is transmitted to the brain. Outer hair cells are a motor structure. Sound energy causes changes in the shape of these cells, which serves to amplify sound vibrations in a frequency specific manner. Lightly resting atop the longest cilia of the inner hair cells is the tectorial membrane, which moves back and forth with each cycle of sound, tilting the cilia, which is what elicits the hair cells' electrical responses.

Inner hair cells, like the photoreceptor cells of the eye, show a graded response, instead of the spikes typical of other neurons. These graded potentials are not bound by the "all or none" properties of an action potential.

At this point, one may ask how such a wiggle of a hair bundle triggers a difference in membrane potential. The current model is that cilia are attached to one another by "tip links", structures which link the tips of one cilium to another. Stretching and compressing, the tip links may open an ion channel and produce the receptor potential in the hair cell. Recently it has been shown that Cadherin-23 CDH23 and Protocadherin-15 PCDH15 are the adhesion molecules associated with these tip links. It is thought that a calcium driven motor causes a shortening of these links to regenerate tensions. This regeneration of tension allows for apprehension of prolonged auditory stimulation.

Neurons

Afferent neurons innervate cochlear inner hair cells, at synapses where the neurotransmitter glutamate communicates signals from the hair cells to the dendrites of the primary auditory neurons.

There are far fewer inner hair cells in the cochlea than afferent nerve fibers – many auditory nerve fibers innervate each hair cell. The neural dendrites belong to neurons of the auditory nerve, which in turn joins the vestibular nerve to form the vestibulocochlear nerve, or cranial nerve number VIII. The region of the basilar membrane supplying the inputs to a particular afferent nerve fibre can be considered to be its receptive field.

Efferent projections from the brain to the cochlea also play a role in the perception of sound, although this is not well understood. Efferent synapses occur on outer hair cells and on afferent (towards the brain) dendrites under inner hair cells

Neuronal Structure

Cochlear Nucleus

The cochlear nucleus is the first site of the neuronal processing of the newly converted "digital" data from the inner ear. In mammals, this region is anatomically and physiologically split into two regions, the dorsal cochlear nucleus (DCN), and ventral cochlear nucleus (VCN). The VCN is further divided by the nerve root into the posteroventral cochlear nucleus (PVCN) and the anteroventral cochlear nucleus (AVCN).

Trapezoid Body

The trapezoid body is a bundle of decussating fibers in the ventral pons that carry information used for binaural computations in the brainstem. Some of these axons come from the cochlear nucleus and cross over to the other side before traveling on to the superior olivary nucleus. This is believed to help with localization of sound.

Superior Olivary Complex

The superior olivary complex is located in the pons, and receives projections predominantly from the ventral cochlear nucleus, although the dorsal cochlear nucleus projects there as well, via the ventral acoustic stria. Within the superior olivary complex lies the lateral superior olive (LSO) and the medial superior olive (MSO). The former is important in detecting interaural level differences while the latter is important in distinguishing interaural time difference.

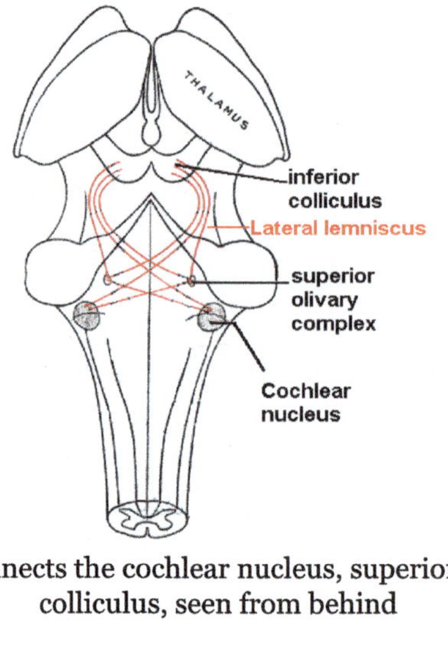

Lateral lemniscus in red, as it connects the cochlear nucleus, superior olivary nucleus and the inferior colliculus, seen from behind

Lateral Lemniscus

The lateral lemniscus is a tract of axons in the brainstem that carries information about sound from the cochlear nucleus to various brainstem nuclei and ultimately the contralateral inferior colliculus of the midbrain.

Inferior Colliculi

The inferior colliculi (IC) are located just below the visual processing centers known as the superior colliculi. The central nucleus of the IC is a nearly obligatory relay in the ascending auditory system, and most likely acts to integrate information (specifically regarding sound source localization from the superior olivary complex and dorsal cochlear nucleus) before sending it to the thalamus and cortex.

Medial Geniculate Nucleus

The medial geniculate nucleus is part of the thalamic relay system.

Primary Auditory Cortex

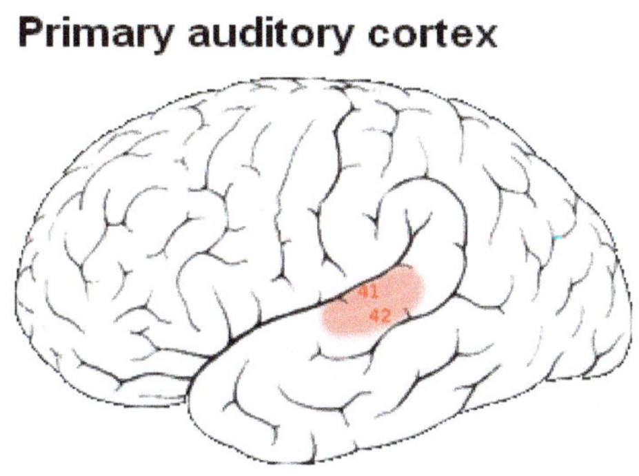

The primary auditory cortex is the first region of cerebral cortex to receive auditory input.

Perception of sound is associated with the left posterior superior temporal gyrus (STG). The superior temporal gyrus contains several important structures of the brain, including Brodmann areas 41 and 42, marking the location of the primary auditory cortex, the cortical region responsible for the sensation of basic characteristics of sound such as pitch and rhythm. We know from work in nonhuman primates that primary auditory cortex can probably itself be divided further into functionally differentiable subregions. The neurons of the primary auditory cortex can be considered to have receptive fields covering a range of auditory frequencies and have selective responses to harmonic pitches. Neurons integrating information from the two ears have receptive fields covering a particular region of auditory space.

The primary auditory cortex is surrounded by secondary auditory cortex, and interconnects with it. These secondary areas interconnect with further processing areas in the superior temporal gyrus, in the dorsal bank of the superior temporal sulcus, and in the frontal lobe. In humans, connections of these regions with the middle temporal gyrus are probably important for speech perception. The frontotemporal system underlying auditory perception allows us to distinguish sounds as speech, music, or noise.

System Overview

The outer ear funnels sound vibrations to the eardrum, increasing the sound pressure in the middle frequency range. The middle-ear ossicles further amplify the vibration pressure roughly 20 times. The base of the stapes couples vibrations into the cochlea via the oval window, which vibrates the perilymph liquid (present throughout the inner ear) and causes the round window to bulb out as the oval window bulges in. Vestibular and tympanic ducts are filled with perilymph, and the smaller cochlear duct between them is filled with endolymph, a fluid with a very different ion concentrations and voltage. Vestibular duct perilymph vibrations bend organ of Corti outer cells (4 lines) causing prestin to be released in cell tips. This causes the cells to be chemically elongated and shrunk (somatic motor), and hair bundles to shift which, in turn, electrically effects the basilar membrane's movement (hair-bundle motor). These motors (outer cells) amplify the perilymph vibrations that initially incited them over 40-fold. Since both motors are chemically driven they are unaffected by the newly amplified vibrations due to recuperation time. The outer hair cells (OHC) are minimally innervated by spiral ganglion in slow (unmyelinated) reciprocal communicative bundles (30+ hairs per nerve fiber); this contrasts inner hair cells (IHC) that have only afferent innervation (30+ nerve fibers per one hair) but are heavily connected. There are 4x more OHC than IHC. The basilar membrane is a wall where the majority of the IHC and OHC sit. Basilar membrane width and stiffness corresponds to the frequencies best sensed by the IHC. At the cochlea base the Basilar is at its narrowest and most stiff (high-frequencies), at the cochlea apex it is at its widest and least stiff (low-frequencies). The tectorial membrane supports the remaining IHC and OHC. Tectorial membrane helps facilitate cochlear amplification by stimulating OHC (direct) and IHC (via endolymph vibrations). Tectorial's width and stiffness parallels Basilar's and similarly aids in frequency differentiation.

The superior olivary complex (SOC), in pons, is the first convergence of the left and right cochlear pulses. SOC has 14 described nuclei; their abbreviation are used here. MSO determines the angle the sound came from by measuring time differences in left and right info. LSO normalizes sound levels between the ears; it uses the sound intensities to help determine sound angle. LSO innervates the IHC. VNTB innervate OHC. MNTB inhibit LSO via glycine. LNTB are glycine-immune, used for fast signalling. DPO are high-frequency and tonotopical. DLPO are low-frequency and tonotopical. VLPO have the same function as DPO, but act in a different area. PVO, CPO, RPO, VMPO, ALPO and SPON (inhibited by glycine) are various signalling and inhibiting nuclei.

The trapezoid body is where most of the cochlear nucleus (CN) fibers decussate (cross left to right and vice versa); this cross aids in sound localization. The CN breaks into ventral (VCN) and dorsal (DCN) regions. The VCN has three nuclei. Bushy cells transmit timing info, their shape averages timing jitters. Stellate (chopper) cells encode sound spectra (peaks and valleys) by spatial neural firing rates based on auditory input strength (rather than frequency). Octopus cells have close to the best temporal precision while firing, they decode the auditory timing code. The DCN has 2 nuclei. DCN also receives info from VCN. Fusiform cells integrate information to determine spectral cues to locations (for example, whether a sound originated from in front or behind). Cochlear nerve fibers (30,000+) each have a most sensitive frequency and respond over a wide range of levels.

Simplified, nerve fibers' signals are transported by bushy cells to the binaural areas in the olivary complex, while signal peaks and valleys are noted by stellate cells, and signal timing is extracted by octopus cells. The lateral lemniscus has three nuclei: dorsal nuclei respond best to bilateral input and have complexity tuned responses; intermediate nuclei have broad tuning responses; and ventral nuclei have broad and moderately complex tuning curves. Ventral nuclei of lateral lemniscus help the inferior colliculus (IC) decode amplitude modulated sounds by giving both phasic and tonic responses (short and long notes, respectively). IC receives inputs not shown, including visual (pretectal area: moves eyes to sound. superior colliculus: orientation and behavior toward objects, as well as eye movements (saccade)) areas, Pons (superior cerebellar peduncle: thalamus to cerebellum connection/hear sound and learn behavioral response), spinal cord (periaqueductal grey: hear sound and instinctually move), and thalamus. The above are what implicate IC in the 'startle response' and ocular reflexes. Beyond multi-sensory integration IC responds to specific amplitude modulation frequencies, allowing for the detection of pitch. IC also determines time differences in binaural hearing. The medial geniculate nucleus divides into ventral (relay and relay-inhibitory cells: frequency, intensity, and binaural info topographically relayed), dorsal (broad and complex tuned nuclei: connection to somatosensory info), and medial (broad, complex, and narrow tuned nuclei: relay intensity and sound duration). The auditory cortex (AC) brings sound into awareness/perception. AC identifies sounds (sound-name recognition) and also identifies the sound's origin location. AC is a topographical frequency map with bundles reacting to different harmonies, timing and pitch. Right-hand-side AC is more sensitive to tonality, left-hand-side AC is more sensitive to minute sequential differences in sound. Rostromedial and ventrolateral prefrontal cortices are involved in activation during tonal space and storing short-term memories, respectively. The Heschl's gyrus/transverse temporal gyrus includes Wernicke's area and functionality, it is heavily involved in emotion-sound, emotion-facial-expression, and sound-memory processes. The entorhinal cortex is the part of the 'hippocampus system' that aids and stores visual and auditory memories. The supramarginal gyrus (SMG) aids in language comprehension and is responsible for compassionate responses. SMG links sounds to words with the angular gyrus and aids in word choice. SMG integrates tactile, visual, and auditory info.

Clinical Significance

- Auditory brainstem response and ABR audiometry test for newborn hearing
- Auditory processing disorder
- Noise health effects
- Tinnitus
- Endaural phenomena
- Gammatone filter – a simple linear model of peripheral auditory filtering

References

- Gelfand, Stanley A. (2009). Essentials of Audiology (3 ed.). New York: Thieme Medical Publishers, Inc. p. ix. ISBN 978-1-60406-044-7. Retrieved 17 March 2015.
- Patrick J. Willems (2004). Genetic hearing loss. CRC Press. pp. 34–. ISBN 978-0-8247-4309-3. Retrieved 23 June 2011.
- Richter, Claus-Peter, and Gulam Emadi. "Tectorial Membrane Stiffness Gradients." Science Direct. 15 Sept. 2007. Web. 2 Apr. 2016.
- Freemana, Dennis M., and C. Cameron Abnete. "Dynamic Material Properties of the Tectorial Membrane: A Summary." Science Direct. June 2003. Web. 02 Apr. 2016.
- Zwislocki, J. J. "Tectorial Membrane: A Possible Sharpening Effect on the Frequency Analysis in the Cochlea." Taylor & Francis. 1979. Web. 02 Apr. 2016.
- Teudt, I. U., and C. P. Richter. "Basilar Membrane and Tectorial Membrane Stiffness in the CBA/CaJ Mouse." Springer. 28 May 2014. Web. 02 Apr. 2016.
- Thompson, A.M., and B.R. Schofield. "Afferent Projections of the Superior Olivary Complex." Wiley Online Library. 15 Nov. 2000. Web. 27 Mar. 2016.
- Janata, Petr, and Jeffry L. Birk. "The Cortical Topography of Tonal Structures Underlying Western Music." Science. 13 Dec. 2002. Web. 27 Mar. 2016.
- Morosan, P., and J. Rademacher. "Human Primary Auditory Cortex: Cytoarchitectonic Subdivisions and Mapping into a Spatial Reference System."Science Direct. Apr. 2001. Web. 27 Mar. 2016.
- Romanski, L. M., and B. Tian. "Dual Streams of Auditory Afferents Target Multiple Domains in the Primate Prefrontal Cortex." Nature.com. Nature Publishing Group, 1999. Web. 27 Mar. 2016.
- Badre, David, and Anthony D. Wagner. "Left Ventrolateral Prefrontal Cortex and the Cognitive Control of Memory." Science Direct. 2007. Web. 27 Mar. 2016.
- Amunts, K., and O. Kedo. "Cytoarchitectonic Mapping of the Human Amygdala, Hippocampal Region an." Springer. Dec. 2005. Web. 27 Mar. 2016.
- Stoeckel, Cornelia, and Patricia M. Gough. "Supramarginal Gyrus Involvement in Visual Word Recognition." Science Direct. Oct. 2009. Web. 27 Mar. 2016.

Permissions

All chapters in this book are published with permission under the Creative Commons Attribution Share Alike License or equivalent. Every chapter published in this book has been scrutinized by our experts. Their significance has been extensively debated. The topics covered herein carry significant information for a comprehensive understanding. They may even be implemented as practical applications or may be referred to as a beginning point for further studies.

We would like to thank the editorial team for lending their expertise to make the book truly unique. They have played a crucial role in the development of this book. Without their invaluable contributions this book wouldn't have been possible. They have made vital efforts to compile up to date information on the varied aspects of this subject to make this book a valuable addition to the collection of many professionals and students.

This book was conceptualized with the vision of imparting up-to-date and integrated information in this field. To ensure the same, a matchless editorial board was set up. Every individual on the board went through rigorous rounds of assessment to prove their worth. After which they invested a large part of their time researching and compiling the most relevant data for our readers.

The editorial board has been involved in producing this book since its inception. They have spent rigorous hours researching and exploring the diverse topics which have resulted in the successful publishing of this book. They have passed on their knowledge of decades through this book. To expedite this challenging task, the publisher supported the team at every step. A small team of assistant editors was also appointed to further simplify the editing procedure and attain best results for the readers.

Apart from the editorial board, the designing team has also invested a significant amount of their time in understanding the subject and creating the most relevant covers. They scrutinized every image to scout for the most suitable representation of the subject and create an appropriate cover for the book.

The publishing team has been an ardent support to the editorial, designing and production team. Their endless efforts to recruit the best for this project, has resulted in the accomplishment of this book. They are a veteran in the field of academics and their pool of knowledge is as vast as their experience in printing. Their expertise and guidance has proved useful at every step. Their uncompromising quality standards have made this book an exceptional effort. Their encouragement from time to time has been an inspiration for everyone.

The publisher and the editorial board hope that this book will prove to be a valuable piece of knowledge for students, practitioners and scholars across the globe.

Index